Implantable Cardiac Devices Technology

David Korpas

Implantable Cardiac Devices Technology

David Korpas
Boston Scientific Czech Republic
Karla Engliše 3219/4, 150 00 Prague
Czech Republic

VSB - Technical University of Ostrava
Faculty of Electrical Engineering and Computer Science
Department of Cybernetics and Biomedical Engineering
17.listopadu 15, 708 33 Ostrava-Poruba
Czech Republic

Acknowledgment: The work and the contributions were supported by the project SP2013/35 "Biomedical engineering systems IX"

ISBN 978-1-4614-6906-3 ISBN 978-1-4614-6907-0 (eBook)
DOI 10.1007/978-1-4614-6907-0
Springer New York Heidelberg Dordrecht London

Library of Congress Control Number: 2013935494

Printed on acid-free paper

Springer is part of Springer Science+Business Media (www.springer.com)

Foreword

Development in the majority of medicine branches is today conditioned by technological advancement. This is also the case with cardiology, where medical devices designed to correct heart rhythm – pacemakers, cardioverters-defibrillators, and biventricular systems – are implanted in order to help a sick heart.

The book *Implantable Cardiac Devices Technology* is targeted at biomedical and clinical engineers, physicians and technicians in practice, students of biomedical disciplines, and all medical staff who are required to understand the basics and details of pacing technology. The book comprises 14 chapters, which are further subdivided according to specific topics. Since readers' level of knowledge concerning the medical part of the issue may differ, chapters dealing with basic heart anatomy, physiology, and arrhythmology are included for the sake of comprehensiveness.

Medical pacing devices are today only developed and produced globally by several producers who, however, make use of certain different technical solutions, algorithms, system parameters, etc. It was our intention to avoid the description of special functions. The book only covers general procedures and parameters common for the systems of all producers.

The book is intended to serve as a monothematic textbook. In order to make the text comprehensible and well arranged for a reader, references to professional literature are only provided once in a respective chapter.

Abbreviations

ABAP	Atrial Blanking post Atrial Pacing
ABAS	Atrial Blanking post Atrial Sensing
AEI	Atrial Escape Interval
AF	Atrial Fibrillation
AGC	Automatic Gain Control
AIMD	Active Implantable Medical Device
AP	Anteroposterior
ARP	Atrial Refractory Period
ATP	Antitachycardia Pacing
ATR	Atrial Tachy Response
AV	Atrioventricular
AVD	Atrioventricular Delay
AVI	Atrioventricular Interval
BOL	Beginning of Life
BOS	Beginning of Service
BPM	Beats per Minute
CI	Coupling Interval
CRM	Cardiac Rhythm Management
CRT	Cardiac Resynchronization Therapy
CRT-D	Cardiac Resynchronization Therapy Defibrillator
CRT-P	Cardiac Resynchronization Therapy Pacemaker
DFT	Defibrillation Threshold
EAS	Electronic Article Surveillance
EF LV	Left Ventricle Ejection Fraction
EGM	Electrogram
EI	Escape Interval
ECG	Electrocardiogram
ELT	Endless Loop Tachycardia
EMI	Electromagnetic Interference
EOL	End of Life
EOS	End of Service
EP	Electrophysiological
ERI	Elective Replacement Indicator
ERN	Elective Replacement Near
ERT	Elective Replacement Time
HRV	Heart Rate Variability
ICD	Implantable Cardioverter-Defibrillator
IM	Myocardial Infarction

LAO Left Anterior Oblique
LRI Lower Rate Interval
LRL Lower Rate Limit
LVBA Left Ventricle Blanking after Atrial Pace
LVEDD Left Ventricular End Diastolic Diameter
LVPP Left Ventricular Protection Period
LVRP Left Ventricular Refractory Period
MPR Maximum Pacing Rate
MS Mode Switch
MSR Maximum Sensor Rate
MTR Maximum Tracking Rate
MV Minute Ventilation
NSR Normal Sinus Rhythm
PAC Premature Atrial Contraction
PAVB Postatrial Ventricular Blanking
PES Programmed Electrical Stimulation
PM Pacemaker
PMT Pacemaker Mediated Tachycardia
pNN50 Percentage of adjacent RR intervals that varied by more than 50 ms
PSA Pacing System Analyzer
PSP Prolonged Service Period
PVAB Postventricular Atrial Blanking
PVARP Postventricular Atrial Refractory Period
PVC Premature Ventricular Contraction
RAO Right Anterior Oblique
rMSSD Root Mean Square of the difference between the coupling intervals of adjacent RR intervals
RRT Recommended Replacement Time
RTTE Radio and Telecommunications Terminal Equipment
RV Right Ventricle
RVC Right Ventricular Coil
RVRP Right Ventricular Refractory Period
SCD Strength-Duration Curve
SDANN Standard Deviation of Averaged Normal R to R intervals
SDI Sensor Driven Interval
SDNN Standard Deviation of all Normal R to R intervals
SQ Subcutaneous
SVC Supraventricular Coil
SVT Supraventricular Tachycardia
TARP Total Atrial Refractory Period
TENS Transcutaneous Electrical Nerve Stimulation
UPR, UR Upper Pacing Rate
USR Upper Sensor Rate
UTR Upper Tracking Rate
VBVP Ventricular Blanking post Ventricular Pacing
VBVS Ventricular Blanking post Ventricular Sensing
VRP Ventricular Refractory Period

Contents

History and Development of Pacing 1

The development of pacing technology has always been closely related to discoveries in the field of electricity and, later, electronic components and materials. The first written records of attempts to pace cardiac nerves or muscles in animals using electric current date back to the end of the eighteenth century [1]. In the nineteenth century, successful resuscitations of patients in cardiac arrest using electric current were documented [2]. In addition, the interest in acupuncture increased; in 1825 electric current was applied for the first time through thin-needle electrodes, derived from acupuncture needles. Thus, electroacupuncture was developed with the purpose of applying electric current to pace muscles, nerves, and organs.

The first attempt to pace the heart using electric impulses was recorded in 1828. Later, experiments on animals were conducted, where cardiac arrest was induced by a chloroform overdose, and the contractions of the heart muscle were restored by means of an electric current. It emerged that the rate of pacing must exceed the intrinsic heart rate to induce the pacing effect. The possibility of inducing ventricular fibrillation using electric current and repetitive cardioversions by strong current pulses was tested [3, 4]. The first portable ambulatory resuscitation apparatus was designed. At the turn of the nineteenth century, discoveries were made in the field of cardiac physiology and in the cardiac conduction system [5]. The cardiac automaticity gradient was discovered, and scientific articles dealing with pathophysiology of tachycardia and bradycardia were published.

It is interesting that the experimental and clinical findings did not result in systematic clinical research in pacing and defibrillation. Research on animals started in Europe only in the 1920s. Since the 1930s, a great number of crucial scientific studies in the field of cardiac electrophysiology was published – in particular in the USA – and today the studies may be retrieved from digitized scientific databases [6, 7].

1.1 The Beginnings of Pacing Technology

The first external pacemakers were produced in the USA in the early 1930s. These devices were operated by a hand crank with a spring motor, which turned a magnet to induce an electric current. The motor was capable of pacing for 6 min. The pulses were supposed to be applied through a transthoracic needle. These pacemakers were named *Hyman I* and *II* after their designer. The devices were too bulky and weighed more than 7 kg. However, type II could be carried in a case with a handle. Later analyses showed that these devices would probably not have been capable of providing effective pacing pulses in real situations, yet they were the oldest known devices designed specifically for the purpose of resuscitation from cardiac arrest. In the 1940s, the first working devices for external defibrillation were described. They made use of alternating current, and their application was successful only in connection with administration of drugs and heart massage [3, 8].

The first implantable pacemaker was implanted in a man in 1958 in Sweden. The pacemaker worked for several hours. The system comprised a steel lead, which was implanted epimyocardially. The pacemaker proper was equipped with a nickel–cadmium battery and sealed in epoxy resin [9, 10].

Because of the insufficient reliability of implantable pacemakers, external pacemakers with a connection to a temporary transvenous lead in the cephalic vein were used for pacing in the late 1950s. In 1959, for instance, a pacemaker intended for a long-term application was used in a 67-year-old patient with second- and third-degree atrioventricular (AV) block. After the implantation of a transvenous lead and subsequent hospitalization, the patient was discharged to home care until November 1962. He was paced by means of a battery-powered device that could also sense the intrinsic

D. Korpas, *Implantable Cardiac Devices Technology*, DOI 10.1007/978-1-4614-6907-0_1,

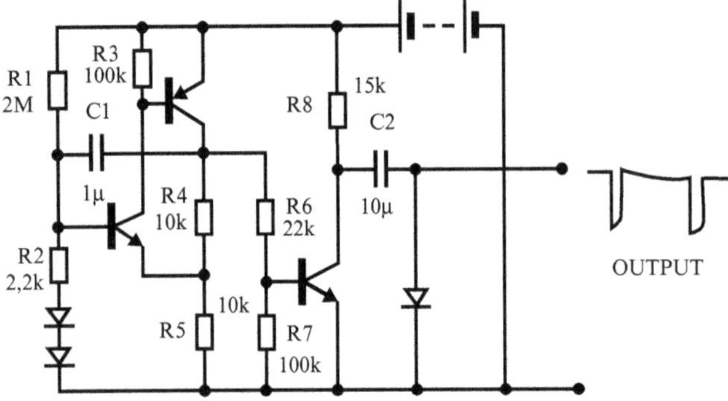

Fig. 1.1 Asynchronous pacemaker circuit (Used with permission of V. Bicik Research Institute for Medical Electronics and Modelling, Prague, Czechoslovakia)

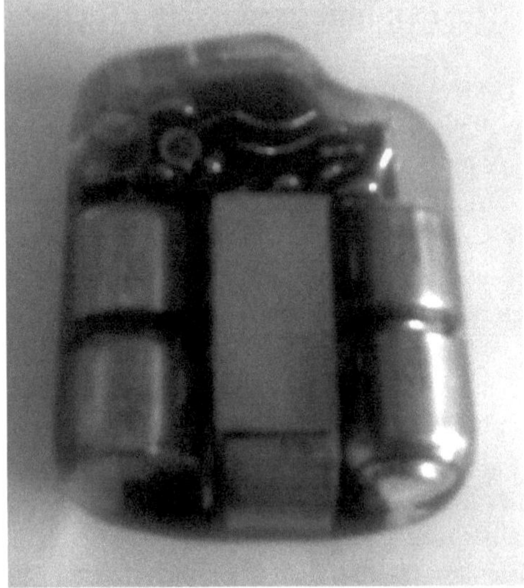

Fig. 1.2 Epoxy resin pacemaker

cardiac activity (but without the possibility of pacing inhibition) and allow modification of the pacing pulse amplitude, measurement of the impedance, etc. The connection of the external device and the implanted lead, however, proved to be problematic because the percutaneous insertion of the lead connected to the external device possessed a risk of infection [10].

First-generation implantable pacemakers only provided asynchronous pacing, that is, they disregarded the intrinsic cardiac activity (Fig. 1.1). The output energy of pacing pulses was higher than required. At that time, pacing at 70–80 pulses per minute, voltage of around 5 V, and pulse width of 1.5 ms was considered appropriate. Epoxy resin was chosen as a biocompatible material with which to plug the seal (Fig. 1.2). The first leads were epimyocardial, when the implantation required a left-sided thoracotomy. The material of the leads was Elgiloy alloy, which the Elgin Watch Company made use of in balance wheels for mechanical watches. In addition, silicon transistors became widespread and enhanced the reliability of circuits. Energy was supplied by electrochemical cells on a zinc–mercury basis. Nickel–cadmium rechargeable batteries also were used [4]. On average, the batteries required replacing after the lapse of a year and a half because of exhaustion. Syncope during second- and third-degree AV block was the main indication that a pacemaker should be implanted.

1.2 Design History

The implantation of the "on-demand" pacing mode was a great step forward, preventing possible competitive pacing that potentially could result in ventricular fibrillation. The first devices were launched in the mid-1960s (Fig. 1.3). The principle was devised by B. V. Berkowitz. In the late 1960s, the first dual-chamber pacing system with the possibility of R wave inhibition was developed. In cases of sinus bradycardia and AV conduction defects, the device paced in the atrium and, after a time lag, in the ventricle. Upon detection of intrinsic ventricle activity, it only paced in the atrium. However, the AV-sequential, dual-chamber pacing mode was not applied until the 1970s.

As a consequence of the introduction of a lithium cell, the dimensions of the device could be reduced. Its electrochemical properties made it easier to estimate the time until battery exhaustion and to schedule device replacement. The most important property of lithium/halogen cells was, however, that they did not produce any gases while being used and could be sealed hermetically. In the mid-1970s, certain manufacturers started using titanium cans instead of the original epoxy resin seal plug. As a consequence of signal processing development, filtration at the input to the sensing circuits was improved, and the impact of electromagnetic interference was minimized [4]. The first noninvasively programmable pacemakers were launched. These devices could be programmed only to a limited extent, providing several options for the adjustment of pacing rate (frequency), pulse amplitude, and, in certain types, sensitivity. Another achievement was the introduction of two-way programmer–implant communication. Nevertheless, fully communicating devices capable of measuring the parameters of a pacing circuit were not produced until the 1980s. The possibility of programming pacing pulse parameters and individual settings of a device in accordance with an individual patient's needs prolonged the longevity of the devices and enabled the treatment of various arrhythmias. In 1974, the first three-position code for designating pacing mode was developed by the Intersociety Commission on Heart Disease Resources. The code corresponds to the first three positions of the code used today. At that time, most pacemakers worked in the VVI mode (Fig. 1.4). Since the beginning of the 1980s, dual-chamber pacing has been applied more often because of technical innovations. The devices were capable of pacing and sensing in both the atrium and the ventricle; naturally, they allowed two-way programmer–device communication and were multiprogrammable.

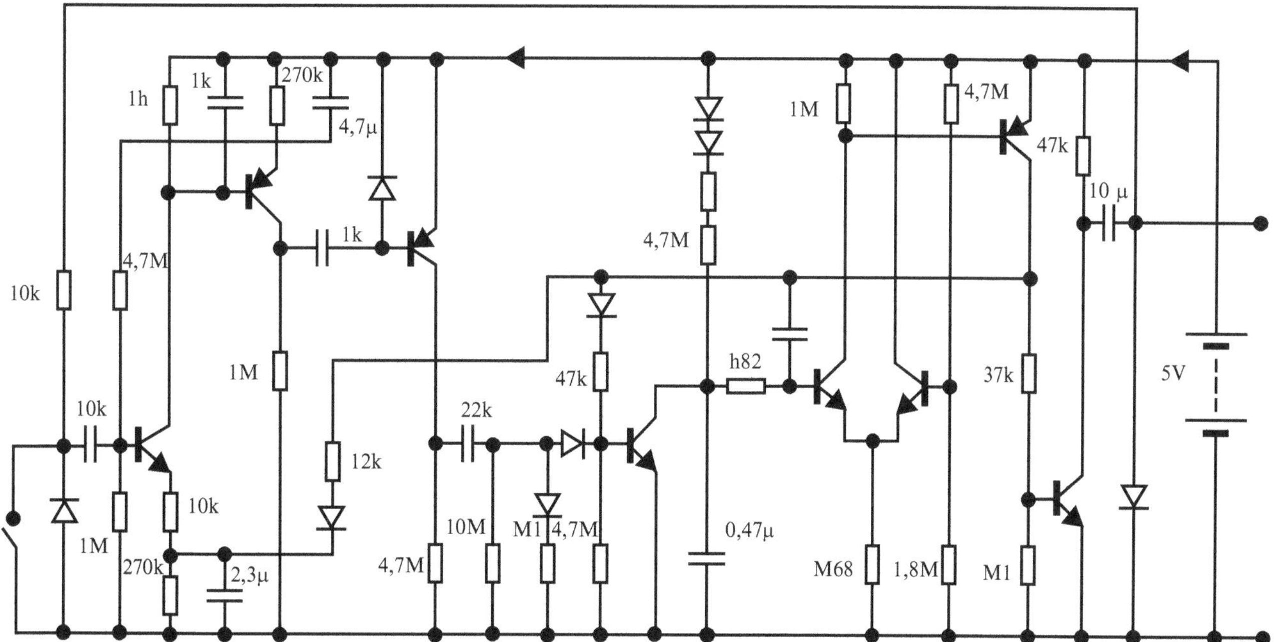

Fig. 1.3 On-demand pacemaker circuit (Used with permission of V. Bicik Research Institute for Medical Electronics and Modelling, Prague, Czechoslovakia)

Fig. 1.4 Pacemaker Tesla LSK 201

The design of a preformed *J*-shaped atrial lead was an important milestone because it facilitated the insertion and fixation of the atrial lead, which had been rather difficult before. In addition to the existing fixation options using tines or a funnel, retractable screw-in active fixation also was tested. Attention was given to the design of pacing electrodes, the use of a rough fractal surface, and the decrease of polarization voltages. Silicone rubber in the lead insulation was replaced with polyurethane; as a consequence, the insertion of two leads into one vein became easier.

The elution of a steroid by the first lead to reduce inflammatory response at the point of contact of the lead and the tissue was put into practice in 1983. In the mid-1980s, pacemakers capable of adjusting pacing based on the patient's activity, sensed by a piezoelectric crystal, were produced. In 1988, the possibility of measuring intrathoracic impedance was introduced, allowing controlled pacing according to a patient's physiological need derived from the respiratory activity. In connection with the development of new therapeutic and diagnostic methods, a new revised five-position code was developed by the Intersociety Commission on Heart Disease Resources. The original code was augmented by the fourth (programmable functions) and fifth (antitachycardia functions) position. The five-position code used today, defined by the North American Society of Pacing and Electrophysiology and the British Pacing and Electrophysiology Group, was approved in 1984, modified in 1987, and further revised in 2002. Important improvements in pacemaker technology are listed in Table 1.1.

The development of implantable defibrillation technology was triggered by the approval of a theoretical concept in 1966. Three years later, an experimental laboratory device was constructed, and the possibility of transvenous defibrillation was tested. In 1975, the first experiment was conducted on a dog. The first defibrillator was implanted in a man in 1980. At first, the devices were implanted in a subcutaneous pocket on the abdomen, and the epimyocardial defibrillation lead was attached above the left ventricle during open thoracotomy. The first implantable cardioverter-defibrillators were equipped only with the function of a shock sent upon the detection of ventricular fibrillation; they lacked any diagnostic functions. The important improvements in implantable cardioverter-defibrillator technology are listed in Table 1.2.

Table 1.1 Pacemaker design milestones [11]

1957	First wearable cardiac pacemaker
1958	First human implant
1960	Fully transistorized pacemaker
1963	Atrial triggered pacemaker (VAT)
1965	On-demand pacemaker patent (concept 1963)
1972	Noninvasive rate and output programmable pulse generator
1973	Lithium-powered, hermetically sealed pulse generator
1976	Long-lasting, lithium-powered pacemaker
1977	Multiprogrammable pulse generators; thin lithium-powered pulse generators
1978	DDD pacemaker (concept 1975)
1980	Pacemaker intervention for supraventricular tachycardia
1981	Rate-responsive pacing; multiprogrammable dual-chamber pacemaker
1998	Automatic capture detection

Table 1.2 Implantable cardioverter-defibrillator (ICD) design milestones [11]

1966	Concept
1969	First experimental model
1969	First transvenous defibrillation
1975	First animal implant
1980	First human implant
1982	Cardioversion available
1985	Food and Drug Administration approval of ICD
1985	Antitachycardia pacing
1986	Transvenous lead
1988	Programmable device
1988	Endocardial implant
1989	Tiered therapy device
1992	Biphasic device
1993	Stored electrograms
1993	Pectoral implants
1994	Active can device
1995	Device-based testing
1996	Dual-chamber device
1997	Dual-chamber, rate-adaptive device
1999	Cardiac resynchronization therapy defibrillator

The first biventricular pacemaker designed for the treatment of cardiac failure was launched in 1995. Devices with a common output channel for the left and right ventricles prevailed in the first decade of clinical application.

1.3 Manufacturing History

During the history of pacemakers, about 40 companies manufactured pacemakers worldwide. During the past two decades, a lot of mergers occurred, so most of the original companies are now part of large multinational corporations such as Boston Scientific (Guidant, Intermedics), ELA (Angeion, Sorin), Medtronic (Vitatron), and St. Jude Medical (Cordis, Pacesetter, Siemens-Elema, Telectronics, Ventritex). Unfortunately, some companies also faded (Omikron Scientific, Tesla). The overview of former and recent pacemaker technology companies is shown in Table 1.3. Because not all historical data are available, the list is maybe not complete. Despite of different technical quality of the implantable systems around the world, their inventors and manufacturers helped a lot of patients suffering especially from advanced or complete heart blocks.

Table 1.3 Overview of pacemakers manufacturers worldwide [12–14]

Country	Company
Brazil	InCor
Canada	National Research Council
China	Quinming
Czechoslovakia	Tesla
England	Geoffrey Davies of Devices
Germany	Biotronik, Cardiotron (formerly GDR)
Italy	Digikon, Medico, ELA-Sorin
India	MediVed, Shree Pacetronix Ltd.
Israel	Omikron Scientific
Netherlands	Vitatron
Russia	Baikal
Sweden	Siemens-Elema
Uruguay	CCC
USA	American Optical, American Pacemaker, American Technology, ARCO Medical Products, Boston Scientific, Cardiac Control Systems, Cardiac Pacemakers, Cook Pacemakers, Coratomic, Cordis, Daig Medcor, Edwards Pacemaker Systems, General Electric, Guidant, Intermedics, Medcor, Medtronic, Pacesetter, Stimulation Technology, St. Jude Medical, Synthemed, Telectronics

1.4 History of Pacing Medical Care

Since the beginning of the 1960s, pacing medical care has developed into one of the biggest and still growing medical device business. The first implant in a given country (see Table 1.4) usually started the pacemaker "rush."

Surveys of pacing practice suggested that new indications accounted for one-quarter to one-half of new implants during the 1970s. Adding new indications for pacemaker implantation meant that the universe of potential patients expanded [15].

Table 1.4 Year of the first pacemaker implant

Year of the first pacemaker implant	Country
1958	Sweden
1960	USA, Uruguay, former USSR (Lithuania), Australia
1961	Germany, Israel, UK
1962	Czechoslovakia, Netherlands
1963	Japan, Poland

2 Basic Principles of Cardiac Pacemaker Technology

2.1 Cardiac Pacemaker Classifications

Nowadays, cardiac pacemaker technology includes a wide range of implantable medical devices, the use of which has increased worldwide. According to valid national and European legislation, cardiac pacemaker technology falls into a group of active implantable medical devices (AIMD). This group is subject to the most severe requirements with regard to safety and reliability.

An up-to-date cardiac pacing system is a medical device that always consists of the main unit itself and between one and three leads, the number of which depends on the type of heart blockage. As a matter of principle, implantable systems might be used within several situations. Treatment of a patient's slow heart rhythm requires implantation of a lead in either the atrium or the ventricle. In the case of absence of the sensed intrinsic heart beat, the pacemaker (PM) sends a stimulus based on defined parameters. With heart blockages of all degrees, one lead senses the contractions of the atria and the second lead initiates the contraction of the ventricle after a delay. The most recent possibility is a solution to ventricular dyssynchrony, created because of structural changes (cardiac failure), by implantation of a third lead epicardially on the left ventricle. If a patient is endangered by a fast heart rhythm (tachycardia), an implantable cardioverter-defibrillator (ICD) is used. Again, it is possible to select only one lead (in the right ventricle) or two leads (one in the atrium and the another in the ventricle). The system with three leads for resynchronization therapy in the case of heart failure can also be used with a defibrillator (a cardiac resynchronization therapy defibrillator [CRT-D]). These devices always are equipped with a sensor for the adaptation of the paced rate (frequency) according to the patient's needs (Fig. 2.1).

2.2 Electric Cardiac Pacing

Cardiac pacing principles are based on the creation of an electrical field between the electrodes and surrounding myocardium by means of an electric stimulus. For the creation of an action potential and its subsequent spontaneous propagation, it is necessary to ensure that a difference of potentials between extracellular and intracellular domains on ectoplasm fall to the value of the threshold potential – from the value of about −80 to about −60 mV. The intracellular domain is charged relatively negatively. The extracellular domain, however, is charged positively. The electrode fixed to the endocardium is compared with practically all cellules in the extracellular domain. The closest surroundings of the electrode are on the same potential as the electrode. So, under an electric stimulus, the extracellular domain is polarized in compliance with the stimulus. The purpose is to induce an action potential on the membranes by changing the electric potential to above the value of the membrane potential. Because the intracellular domain is charged relatively negatively, a decrease in potential of the membrane can be achieved by decreasing the potential of the extracellular domain using a negative pulse. A positive electric stimulus can also be used for pacing. However, the amplitude must be a little higher. The action potential created on the cellules depolarized directly by the electrode is propagated by biophysical mechanisms on the surrounding cellules, and they also are depolarized. The highest current density from the electrodes is at the edge between the electrode and the tissue. It decreases as the distance from the electrode increases.

A minimal value of a certain physical quantity at which a consistent cardiac depolarization has been safely created and propagated is called a pacing threshold. The pacing

D. Korpas, *Implantable Cardiac Devices Technology*, DOI 10.1007/978-1-4614-6907-0_2,

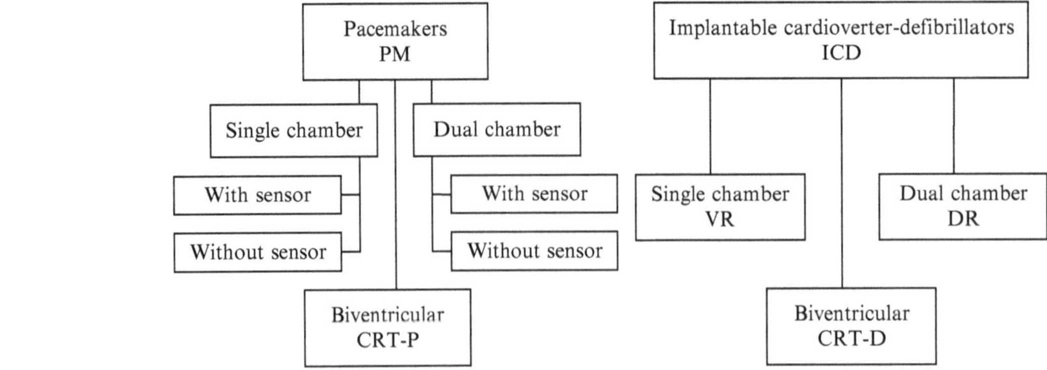

Fig. 2.1 Division of implantable cardiac devices

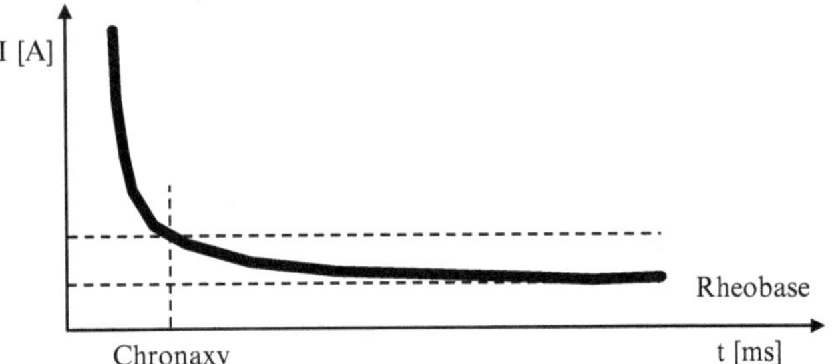

Fig. 2.2 Strength–duration curve

threshold can be expressed in terms of amplitude, pulse width, or energy according to the direct proportion $E \approx U^2 * t$. Cardiac pacing uses rectangular electric stimuli with programmable amplitude and width. Excitability of the heart muscle can be expressed by parameters of a cardiac electric stimulus that is able to activate cardiac depolarization, and the so called Hoorweg-Weiss curve (strength–duration curve) that expresses a relation between current amplitude and pulse width of the pacing threshold is used for this (Fig. 2.2). The curve has the shape of hyperbola, and there are two characteristic values that are defined on it:

- Rheobase – a minimal pacing threshold current for the theoretically infinite width of the pulse
- Chronaxy – a pulse width at which the pacing threshold is equal to twice the rheobase. In practice, within the rheobase definition, the infinite width of the pulse is substituted by a definite one, for example 2.0 ms.

Either a doubled value of the voltage threshold or a tripled value of the pacing threshold width is considered to be a safe reserve of an electric stimulus output. The issue of the pacing threshold is much more complicated and complex. The pacing threshold is influenced by, for example, the type and material of the lead used, by the distance between the electrodes, and by the state of the tissue. The pacing threshold also changes within the time after a lead implantation. However, a considerable increase has not been observed yet, thanks to the use of steroids. Within a daily cycle, the pacing threshold is higher during sleep; it falls during the waking state and decreases even more distinctly during physical exertion. The pacing threshold is also influenced by pharmaceuticals: it rises especially after the use of β-blockers and class I antiarrhythmics but falls after corticosteroid use. Occasionally, a brisk and inexplicable rise of the pacing threshold of some patients can be observed. This is designated as an exit block.

2.3 Energy Sources and Longevity of Implantable Devices

A source of energy was always considered an important problem of AIMDs. It is necessary to ensure reasonable energetic capacity and reliability and to further their operational performance of characteristics such as the voltage, self-discharge current, energy density per volume unit, biological compatibility, and structural shape. Historically, energy sources can be divided into three groups: electrochemical, radioisotopic, and biological sources. The electrochemical (galvanic) cells comply best with the requirements stated above, excluding the capacity. Their output voltage is not dependent on the output, and they have a good structural formability. At the end of the 1960s, radioisotopic thermoelectric generators were applied. They had much better energy capacity – they could operate for longer than 30 years. However, a high price and possible danger of radioactive substance leakage were disadvantages. Either biogalvanic cells, which operate with body fluids such as electrolytes during electrochemical reactions, or metal/oxygen biofuel cells, in which the metal anode is consumed by oxidative corrosion and the cathode decreases the oxygen present in body fluids, have been used as biological or biochemical power sources. However, undesirable reactions of the tissue were observed in these cases. Electromechanical converters can also be classified among biological sources, but they required the patient's movement to make them work.

Nowadays, energy sources for implantable devices include monocell and polycell lithium–iodine batteries. Voltage of

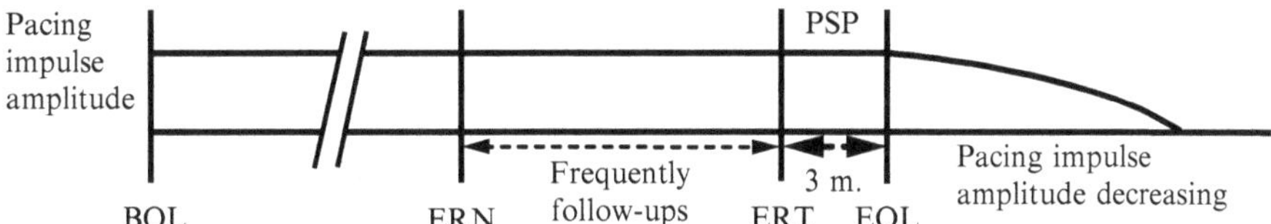

Fig. 2.3 Longevity phases [32] (© 2012 Boston Scientific Corporation or its affiliates. All rights reserved. Used with permission of Boston Scientific Corporation)

such a battery is always about 3 V, which is given by the electrochemical potentials of lithium and iodine. The capacity, which depends on the type of battery, ranges from 0.8 to 1.8 Ah or more. The current consumption is approximately 10 μA; for example, for a particular type of pacemaker, current consumption amounts to 13.3 μA during pacing and 10.3 μA during inhibition [16]. The life span of a device depends on the pacing mode used and the number of electric stimuli; the lifespan of defibrillators especially depends on the number of shocks delivered or charging cycles. After discharging the battery, a new device must be implanted. The leads usually remain until they are damaged.

Pacemaker battery status can be evaluated either by means of a telemetric connection using a programmer or by the output paced rate when a magnet is positioned over the pacemaker. A window on the programmer showing the battery status might display a date of the last battery test, previous and recent indicators of the battery's status, the recent output of the pacemaker when a magnet is used, as well as estimated service time remaining based on the measurements indicating the rest of the battery capacity. A valid technical standard [17] requires an AIMD containing a source of energy must provide a warning signal on depletion of the energy source in advance. Time period of the warning under normal usage of the device must be at least as long as the recommended time period between the clinical ambulatory follow-ups. The manufacturer designates when an exchange of the device is recommended. The standard [17] defines the following stages of service life (Fig. 2.3) according to the remaining electric capacity of batteries:

- Beginning of service (beginning of life) – the implantable device is authorized by the manufacturer for the first time as capable of launching.
- Recommended replacement time (elective replacement time; elective replacement indicator) – indicator of the energy source reaches a specific value that had been appointed in advance by the manufacturer of the implantable device for its recommended exchange. This point is the beginning of the prolonged service period.
- Prolonged Service Period– the time period after the point of the recommended replacement time when the implantable device continues to operate as specified by the manufacturer.
- End of Service (end of life [EOL]) – the prolonged service period has expired and another pacing function is not specified nor can be expected.

Table 2.1 Usual magnet output pacing rates [16, 18]

Manufacturer	Beginning of Life (per min)	Elective replacement time (per min)
Biotronik	90	80
Boston Scientific	100	85
ELA	96	80
Medtronic	85	65
St. Jude Medical	98.6	86.3
Vitatron	100	86

Of the terms and abbreviations above, the standards use the first ones listed. However, in practice, those in parentheses are used more often. Some manufacturers might also use an identifier called *elective replacement near* (ERN). After this point, it is recommended that patient follow-up be performed more often.

Approximately 3 months after elective replacement time, when the battery is being gradually discharged, the device reaches the stadium called the end of life (EOL). When the EOL stadium is reached, some arrangements dealing with maximal reduction of the power consumption are made automatically. The mode of dual-chambered pacemakers changes to a single-chamber mode (DDD and VDD changes to VVI) and the lower rate limit decreases. With further gradual discharging of the battery during the EOL state, the pacemaker reduces the amplitude of electric output. When the EOL is reached, the telemetry does not have to be guaranteed any longer.

If a magnet with the appropriate features is positioned over the implanted pacemaker (it is not applied for defibrillators) and if this function has not been changed by programming, the mode of pacing changes from the programmed mode to an asynchronous mode (D00, V00, or A00), and the paced rate (frequency) is set according to the manufacturer's requirements (Table 2.1). That way, it is possible to check the battery status of the implanted pacemaker if a programmer is not available.

2.4 X-Ray Identifier

According to the standards [17], in case of unexpected change in performance, an implantable device must be identifiable by a noninvasive procedure that does not require the use of tools that are usually unavailable at hospitals. Specific devices (e.g., a programmer) are considered unacceptable. Therefore, pacemakers and defibrillators are provided with an identifier located on the device head that is visible on a radiographic image or under a skiascope. This identifier serves as noninvasive confirmation of the manufacturer and the necessary applications of the programmer. Older devices used to be marked with a numeric code. Nowadays, codes using letters to identify manufacturers are used more often, together with numeric identification of the necessary software of the programmer or identification of the determined pacing mode.

2.5 Programmer Usage

Communication with an implanted device is realized by means of a specialized device called the programmer. It enables programming of parameters of electric stimuli, measurement of sensed signals of the heart's beat and electrical features of the system, and selection of all other parameters. Data transfer between the programmer and the device occurs by means of telemetric inductive coupling or a wireless radio signal.

Programmers are computerized devices with their own operating system. They contain general service software and special applications for particular devices or groups of devices. The user interface is a touch screen that includes buttons for sending data to the implanted device, for diverting the therapy, or for selection of paper advance speed in a case of electrogram (EGM) or electrocardiogram (ECG) records. Furthermore, the built-in parts include an ECG monitor; an internal printer; a device for data disc input (diskettes, flash drive); slots for connecting an external printer, monitor, or keyboard; and inputs from electrophysiological monitoring systems.

On the main screen of the system, cross-referenced entries are available, enabling access to information on the set functions before input to application software, as well as language selection, an easy-to-use diagnostic ECG monitor, a mode for a quick automatic interrogation of the device, and others are available. Regarding the ECG monitor, it is possible to change the speed of motion, to set an input amplifier or an input filter of the surface ECG, or to display electric stimuli spikes. Identification of the implanted device by means of a telemetric sensor positioned over the pacemaker and downloading data from it is designated as reading, or *interrogation*. The interrogation is the first step of all sessions during follow-up. At the initial interrogation, the information on, for example, the parameter settings, patient data, diagnostic data, and battery status, are copied from the device memory.

Parameter values might be changed by touching the pointer to the appropriate parameter window and by lifting the pointer off the screen. After execution of changes to parameter values, the change will appear in the window until it is programmed into the device. After a new parameter is defined, its interactions with the other parameters are evaluated immediately. If a new value breaks the limits of interaction within the application, an icon will appear that reports parameter failure. The failure is described in the interaction window and a solution is proposed. To continue programming changes, it is necessary to make a correction of the influenced parameter first. If it is necessary to make a permanent record of the data interrogated at a follow-up, the programmer offers printed reports containing actual values of parameters, data on therapy history, information on the device's battery status, and programmed data about a patient (date of implantation, type of leads, indication for device implantation, etc.).

2.6 Magnet Usage

Because of the possibility of an emergency effect on the behavior of the implantable devices when a programmer is unavailable, the implantable devices are equipped with a magnetic switch (called a reed switch). Technically, it deals with a reed relay. Its contacts usually are disconnected in a resting state. For making contact with this relay, a magnet with induction of more than 1 mT is used. According to the valid standards [17], the devices must be resistant to magnetic fields up to 1 mT. In practice, small permanent magnets in the shape of a horseshoe, an annular ring, or a prism are used. Some manufacturers supply a magnet as a part of the telemetric wand of the programmer. In general, every pacemaker has a designated response when a magnet is positioned over the device. For pacemakers, it deals with switching to the asynchronous mode and pacing using a defined paced rate according to the battery status. Because the magnet switches off the sensing input amplifier, in this way it is possible, for example, to interrupt pacemaker-mediated tachycardia. Regarding defibrillators, it deals with elimination of tachycardia therapy (shocks or antitachycardia pacing). After the magnet is lifted off the device – after repeated disconnection of the reed relay – the device returns to its normal, originally programmed mode.

A setting determining the response to a magnet might be programmed, depending on the type of the device and the manufacturer. For example, the following settings of defibrillators are available:

- Off (no response when the magnet is positioned over the device),
- Save EGM (it saves an actual EGM),
- Inhibit tachytherapy (therapy application is stopped; or defibrillator mode is switched over).

Some previous systems were equipped with certain possibilities for the measurement of pacing threshold during application of a magnet. For example, the first three pulses were asynchronous, with a paced rate of 100 pulses/min. They were followed by asynchronous pacing at a programmed paced rate. The first and second electric stimuli had the programmed width, whereas the third one had only 75 % of the programmed width. Loss of the paced rate with the third stimulus meant a small safety reserve. Another system applied 16 asynchronous stimuli with a paced rate of

100 pulses/min followed by 16 stimuli with a paced rate of 125 pulses/min. During this faster pacing, the output voltage of the stimulus gradually decreased to zero. However, after removal of the magnet, the output returned to the programmed value. If an implanted device is exposed to a strong, external, static magnetic field, unwanted contact with the magnetic switch poses a considerable danger, especially in the case of defibrillators, when the therapy would be suppressed. The reed relay is a mechanical part only. Its reliability is lower than that of electronic systems. Therefore, a magnet should be used with the highest caution.

2.7 Implantable Systems Compatible with Magnetic Resonance Imaging

Originally, patients with an implanted pacemaker system were not allowed to be imaged using magnetic resonance (MRI). Nowadays, some innovations have occurred in this area [19]. By launching into clinical practice a pacemaker compatible with MRI together with leads compatible with MRI, the biggest contraindication of a magnetic resonance procedure was solved [20].

Devices incompatible with MRI might cause interference in the surgical implant by a static magnetic field, the gradient of the magnetic field, electromagnetic waves, or a combination of these phenomena. Potentially, they might cause vibrations; activate tensile and twisting forces; make contact with the magnetic switch; or cause electromagnetic interferences, failures of cardiac pacing, changes of programmed parameters, or destruction of electronic circuits.

Changes to the construction and programming of systems compatible with MRI constrain the possibility of the phenomena stated above. To minimize the power that is induced on a lead, the capacity at the lead's input to the device has been changed. A classic magnetic switch that could have made contact by the impact of a direct current magnetic field was replaced by a Hall sensor. Because of the influence of attraction forces caused by a strong direct-current magnetic field, usage of ferromagnetic parts was constrained considerably. Additional protection of the internal feeding circuit forestalls the power induced in the loop, avoiding the disturbance of the feeding circuits by telemetry. From the perspective of programming a special cardiac pacing mode, asynchronous programming often is introduced. Furthermore, the collection of diagnostic data and therapy of atrial arrhythmias is interrupted. The construction and internal arrangement of leads have been changed (number of conductors, gradient, diameter, etc.) so that the interactions with gradient magnetic field were eliminated and electrode heating was reduced.

Pacemakers are not automatically compatible with MRI. First of all, before the imaging, it is necessary to program the pacemaker to the MRI Safe mode; after the radiological procedure and during consequent follow-up, it must be reset to the current mode repeatedly. The cardiac pacing systems currently approved as MRI compatible have certain limitations, for example they are compatible only with closed types of MRI devices with an external field of 1.5 T or with changes in amplitude gradient up to 200 T/m/s. That a pacemaker was authorized for 1.5T devices means that it must not be used for devices with either higher or lower external magnetic fields [20].

2.8 Device Construction and Materials

Materials used for the construction of implantable devices must be biologically inert, nontoxic, sterilized, and capable of long-term immunity against conditions in an organism's internal environment. All parts of the implantable systems, including the electronics and leads, must be produced from biocompatible materials.

Solution of circuits uses custom-made microprocessors or microcontrollers using complementary metal oxide semiconductor technology. For controlling the input and output or program settings, read-only memory with a capacity of 1–2 kB and a word width of 8–32 bites are used. Long-term diagnostic data, EGM, or sensor control output is saved to random-access memory. Growing diagnostic possibilities of the devices put more demanding requirements on the random-access memory capacity.

From a mechanical point of view, the implantable devices consist of a case and a header. The case is made of titanium or a titanium alloy and contains all the electronics, battery, capacitors, and output circuits (Fig. 2.4). Data dealing with

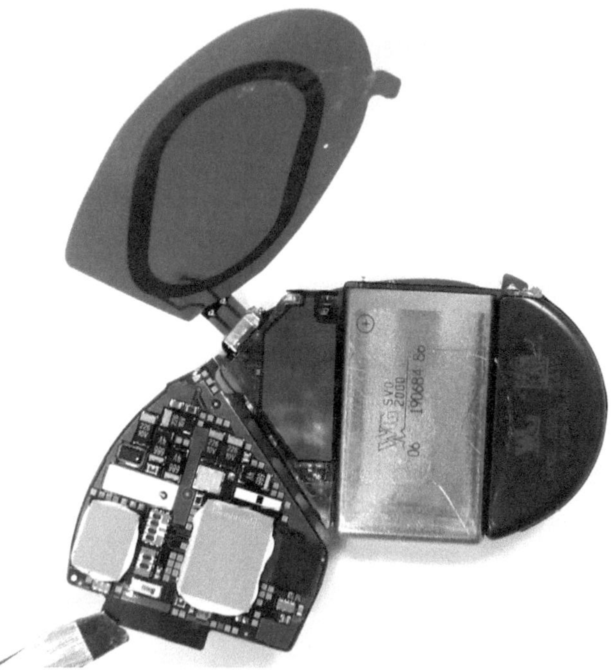

Fig. 2.4 ICD (inside view)

the manufacturer and type of device, serial number, and configuration of lead connections are stated on the device case. In addition, the wand that serves as the connection for the implantable leads coming from the heart is positioned here. Conductors from the input and output elements and leads are led to the header, where they are fixed by setscrews. Every pacing or shock electrode has a contact to which it is fixed by a setscrew or spring. To ensure the device is waterproof, the setscrews are covered by a seal plug. To tighten the setscrews, a bidirectional torque wrench (screwdriver) is used. A correct pressing force of the setscrew on the electrode and protection against damaging by overtightening is ensured by the torque wrench calibration. If a certain torque was exceeded, a handle starts audible skipping and higher torque is no longer generated. Nowadays, a torque wrench no. 2 with a hexagonal tip of about 0.9 mm (0.035 in.) is used. In spite of that, it is recommended that the torque wrench supplied by the manufacturer be used for every device because compatibility of the torque cannot be ensured. In exceptional cases of exchanging a device with a discharged battery, it is also possible to use a fixed, nontorque spanner of the given size for disconnection instead of the torque wrench.

3 Heart Anatomy and Physiology

To provide comprehensive information on pacing and defibrillation methods, the fundamentals of the anatomy and physiology of the heart need to be discussed. An overview of cardiac rhythm disorders and their pharmacological management will follow in Chap. 4. In terms of the pacing method, one has to be familiar with the general structure of the heart and, in more detail, with the right atrium and the right ventricle, where endocardial leads are placed. It is also necessary to understand the anatomy of the conduction system and coronary veins as well as imaging projections.

The basis of the heart's electrical activity is the action potential of cardiac cells. There are two types of cardiac cells: cells of the conduction system and myocardial contractile cells. The function of the cells of the conduction system is to conduct impulses to myocardial contractile cells, which contract to pump blood. They also differ in how the action potential passes along their membranes. For optimal cardiac output, and thus blood supply to the whole body, it is essential to ensure not only contractions of the heart chambers, but also their correct timing. The purpose of pacing is to cause, by supplying external electrical energy, the generation of an action potential in the immediate surroundings of the pacing electrode that is further propagated by biophysical mechanisms.

3.1 Heart Anatomy

The heart is a hollow muscular organ that rhythmically contracts to push blood through the bloodstream [21]. The heart of an adult weighs approximately 230–340 g; the weight of the heart in women is, on average, about 15 % lower than that in men. The weight of the heart depends on the volume of the heart muscle, which varies individually. The average dimensions of the heart are $13 \times 9 \times 6$ cm. The heart is situated behind the sternum, in the mediastinum, with one third located to the right of the middle and two thirds located to the left. It lies in the pericardium, which encloses it.

The external shape of the heart resembles an inverted irregular cone (Fig. 3.1). Its apex is directed downward and forward, its base backward and upward. At the base, where the atria are found, large veins enter and arteries – the aorta and the pulmonary artery – exit. An appendage arises from the edge of each atrium. This is the site where endocardial atrial pacing leads are attached. The right and left ventricles extend from the atria to the apex of the heart. Under physiological conditions, the left ventricle is larger and has a stronger wall than the right one. This results from the physiological difference in their functions: the greater circulation powered by the left ventricle requires higher pressure. Each ventricle has an inflow part and an outflow part.

The right atrium receives deoxygenated blood from the greater circulation. The superior vena cava enters posteriorly and superiorly; the inferior vena cava enters from the left and inferiorly through the diaphragm. By means of internal structures, the flow of blood is directed in such a manner so as to prevent turbulent flow. The coronary sinus, where the coronary veins enter, is located at the posterior atrial wall. This is where pacing leads are implanted in the left ventricle. This orifice is partly covered by the semilunar valve. The left atrium is separated from the right atrium by the atrial septum. Toward the apex, the right atrium meets the right ventricle, and the orifice between them is guarded by the tricuspid valve. During systole, cusps of this valve close together, preventing regurgitation from the right ventricle into the right atrium. The walls of the inflow part of the right ventricle are covered by muscular trabeculae and elongated ridges. In contrast, the outflow part of the right ventricle is smooth. The orifice of the outflow tract is guarded by the pulmonary valve. The left atrium begins where the pulmonary veins empty at its posterior wall. The walls are approximately 3 mm thicker than those of the right atrium, and they are smooth. In a downward and forward direction, the left atrium opens into the left ventricle through the bicuspid (mitral) valve. The inflow part of the left atrium is larger than that of the right atrium, whereas the outflow part is shorter. The right and left ventricles are separated by the ventricular (interventricular) septum. It is as thick as the whole left ventricular wall and arches into the right ventricle.

D. Korpas, *Implantable Cardiac Devices Technology*, DOI 10.1007/978-1-4614-6907-0_3,

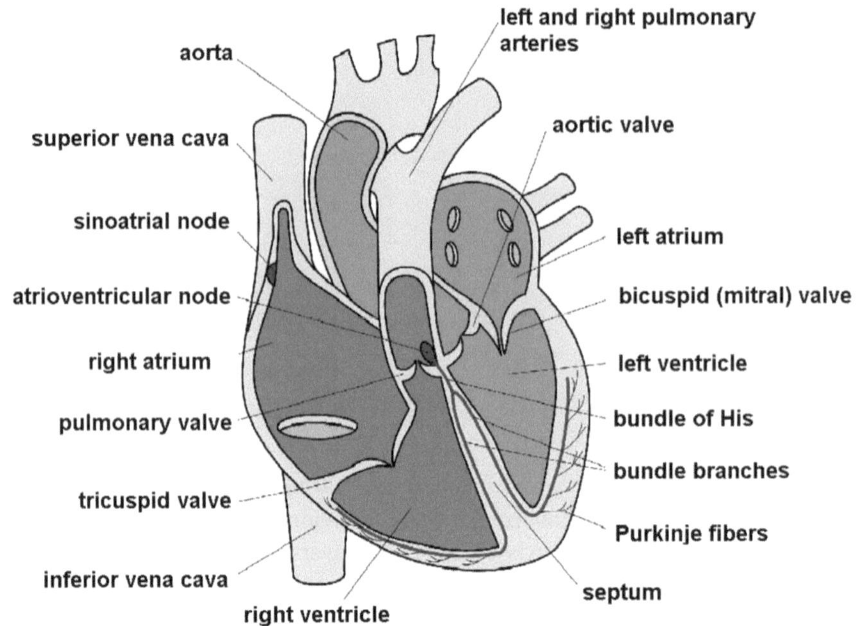

Fig. 3.1 Heart cross section and conduction system

3.2 Structure of the Heart Wall

The wall of the heart is composed of three layers: the epicardium, the myocardium, and the endocardium (Fig. 3.2). The epicardium is a serous outer sac of the heart wall overlying a thin layer of elastic fibrous tissue, which attaches the epicardium to the myocardium. Here, fatty tissue is found in places, particularly in the depressions along the superficial cardiac arteries, veins, and nerves.

The myocardium is a muscular layer made up of a special type of striated heart muscle. It is the main component of the heart wall and has the greatest thickness. It consists of fibers and individual cells connected in a spatial network. Myocardial cells have an oval nucleus surrounded by contractile myofibrils that have a structure similar to that of skeletal muscle fibers. The myocardial layer in the atrial walls and septum is much thinner than that in the ventricles. There is more connective tissue between muscle stripes in the atrial myocardium than in the ventricles. The muscle of the left ventricular wall and septum is about three times as thick as that of the right ventricle. The atrial myocardium is composed of two layers: deep and superficial. Arches and rings of the deep layer encircle each of the atria separately. The superficial layer forms longer transverse stripes that pass from one atrium to the other. The ventricular myocardium has three layers that are mutually intertwined to form a common system.

The endocardium is an intracardial membrane that lines the cardiac cavity. It is smooth, transparent, and glistening. It consists of a single layer of flat endothelial cells overlying connective tissue with collagen and elastic fibers. Elastic fibers are more abundant in the atria than in the ventricles. There are smooth muscle cells in the connective tissue layer; these are more abundant in the atrial and ventricular septa. Endocardial thickness is 50–200 mm. Where thicker, the endocardium is a whitish color; where thinner, the myocardial muscle is visible.

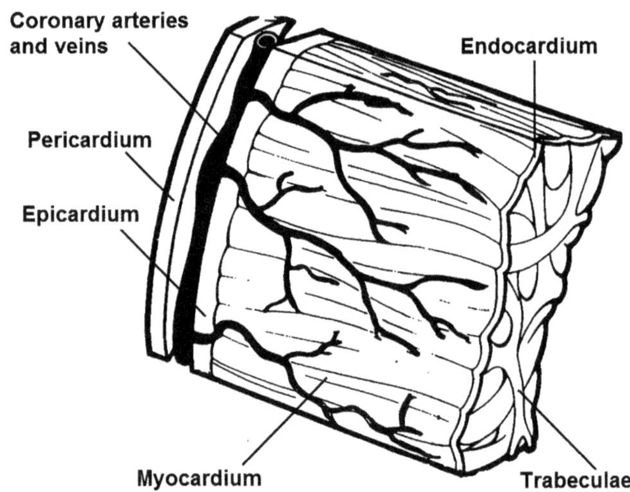

Fig. 3.2 Cardiac walls and coverings

3.3 The Conduction System

The conduction system is a network of specialized cellular structures that generate and conduct impulses. The action of the myocardium is not dependent on additional innervation, and nerves coming to the heart only accelerate or slow down the automatic action. Thus, the myocardium itself is a source of electrical signals required for depolarization of myocardial contractile cells. Cells of the conduction system have larger and more rounded nuclei than myocardial contractile cells. The conduction system (Fig. 3.1) consists of the sinoatrial node, internodal pathways, the atrioventricular node, the atrioventricular bundle of His, right and left bundle branches, and Purkinje fibers.

The principal part of the conduction system is the sinoatrial (SA) node. This is where electric impulses that determine the heart's rate are primarily generated. The SA node is an elongated structure, 10–20 mm long, about 3 mm wide, and 1 mm thick. It is located in the wall of the right atrium beside the opening of the superior vena cava near the upper border of the atrial appendage. It is situated deep within the myocardium, approximately 1 mm from both the endocardium and the epicardium. An impulse arises in central nodal myocytes and is transmitted by slow myocyte-to-myocyte conduction to the subsequent segments of the conduction system.

The atrioventricular (AV) node is a structure located at the boundary between the atria and the ventricles, approximately 10 mm from the opening of the coronary sinus. It is roughly oval in shape, 7–8 mm long, about 1 mm wide, and 3 mm thick. It is made up of myocytes with a network of collagen fibers. Its purpose is to delay the conduction of impulses, which is referred to as AV delay and is thought to be achieved by accumulating a large number of myocytes with a slow conduction velocity. An impulse from the SA node is conducted to the AV node via internodal pathways. They go through the right atrial wall and enter the AV node or as far as the AV bundle. These pathways include the anterior internodal pathway, middle internodal pathway, posterior internodal pathway, interatrial bundle, collateral fibers, and accessory AV bundles.

Arising from the anterior margin of the AV node is the AV bundle of His. It is a compact band of myocytes surrounded by a layer of vascularized connective tissue. At transition to the muscular part of the septum, it divides into the right and left bundle branches. The right bundle branch is a slender, round fascicle that passes down on the right side of the septum toward the apex. It first runs in the myocardium, then below the endocardium, and it continues to divide, forming numerous Purkinje fibers. The left bundle branch is composed of many fine fascicles with a fibrous sheath.

The fascicles gradually divide toward the apex, where they branch along the ventricular walls into Purkinje fibers composed of typical large myocytes. Branching of these fibers proceeds proximally from the apex of the heart. As a consequence, contraction of the ventricular myocardium also proceeds from the apex to the base of the heart as well as from the internal (the endocardium) to the external layer.

3.4 Heart Vessels

Oxygenated blood is brought to the heart walls by the right and left coronary arteries that arise immediately at the origin of the aorta. Deoxygenated blood is removed from the heart walls by cardiac veins. With respect to the pacing method, the anatomy of the coronary arteries is of minor importance only. However, the anatomy of the cardiac veins, into which left ventricular pacing leads are placed, is important.

The coronary sinus that empties posteriorly into the right atrium is the main venous outflow. The great cardiac vein that begins at the apex of the heart and encircles the left heart empties into the coronary sinus. It drains blood from the left and right anterior sides of the left ventricle. The posterior left ventricular vein that collects blood from the posterior wall of the left ventricle empties into it at the posterior side. The middle cardiac vein passes from the apex of the heart directly to the coronary sulcus and drains into the coronary sinus near its entry into the right atrium. The cardiac veins contain no valves. However, valves sometimes do occur at the openings of the major veins into the coronary sinus. The veins contain numerous anastomoses, sometimes even extracardial ones.

3.5 X-Ray Projections of the Heart

Pacing and defibrillation leads are implanted under X-ray skiascopic guidance. For this reason, it is advisable to be familiar with the basic projections used in the imaging of the heart and cardiac veins.

When implanting leads into the right-sided heart chambers, the heart is most commonly viewed in the anteroposterior (AP) projection. In the case of a venogram and left ventricular lead placement, oblique projections are used: right (right anterior oblique) and left (left anterior oblique) under an angle of 30°–60°. On an AP projection, margins of the cardiac shadow are identifiable. On oblique projections, dimensions of the heart chambers are visible and, after obtaining a venogram, a clear image of the left ventricular venous bed is available on a left anterior oblique projection.

3.6 Cell Electrophysiology

The cell body is enclosed by the cell membrane. Functionally, the membrane is an important cell organelle involved in maintaining the intracellular environment and composing the extracellular environment.

The intracellular and extracellular environments are made up of electrolytes, a solution of various concentrations of ions [22]. The predominant intracellular ion is K^+, having a

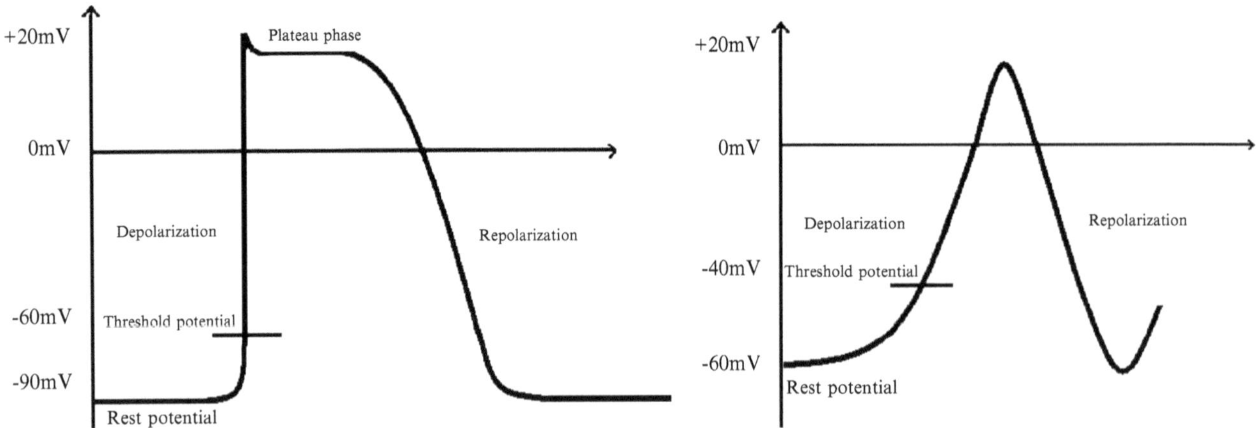

Fig. 3.3 Membrane action potential – muscle cells and cells of the conducting system

concentration about 30 times higher inside than outside the cell. The cell membrane is readily permeable to a potassium ion. The predominant extracellular ion is Na^+, to which the cell membrane is only slightly permeable. Indeed, the membrane has a specific permeability to individual ions, for example, K^+, Na^+, and Cl^- ions or protein anions. This permeability of ion channels is mainly determined by the intracellular concentration of Ca^{2+} ions.

The system strives to create a thermodynamic balance by equalizing ion concentrations on both sides of the membrane. Ions that pass through the membrane thus diffuse into the space where their concentration is lower. However, oppositely charged ions do not diffuse because the membrane is not permeable to them. Therefore, diffusion of ions is halted by an electrical field that occurs as a result of separation of positive and negative ions by the membrane. The voltage occurring at the cell membrane between intracellular and extracellular environments is referred to as the resting membrane potential. Ions thus indirectly affect the magnitude of the potential. However, it is actually electric potential difference, that is, voltage. The magnitude of this voltage ranges from a single to hundreds of millivolts. The extracellular space is positive, whereas the intracellular space is negative. For myocardial contractile cells, the typical resting membrane potential is –90 mV; in the sinoatrial node it is only about –45 mV.

The action potential is a rapid change in the electric potential on the membrane of some cells. Within milliseconds, the intracellular space increases from a value of –90 mV to that of +20 to +30 mV. In excitable membranes, this change in the potential is propagated to the surroundings. An action potential can be generated by chemical processes, external phenomena, arrival of an impulse, or a change in the potential of the membrane – in general, by any phenomenon that lowers the resting membrane potential of a given cell to a threshold level. When the resting membrane potential is negative, this threshold level is at an absolute value of approximately 20 mV higher (closer to zero).

The process of action potential begins by the opening of sodium channels, which results in a rapid rise of the potential into positive values. This phase, accompanied by an influx of Na^+ ions into the cell, is referred to as depolarization. Almost simultaneously, the permeability of potassium channels is increased and potassium ions flow out of the cell. At that point, the rise of the potential is stopped and its subsequent decrease occurs. This phase is called repolarization. From the beginning of depolarization through approximately two thirds of repolarization, the membrane is in an absolute refractory period, that is, it is unexcitable. It cannot be depolarized again, not even with an intensive stimulus, because most of the sodium channels are inactive. The channels cannot open until the membrane voltage returns to a value of around –40 mV. After approximately two thirds of the repolarization phase, depolarization can be evoked again with an above-threshold stimulus; this phase is referred to as the relative refractory period.

The membrane temporal and voltage parameters are dependent on the cell type and are shown in Fig. 3.3. In myocardial contractile cells, the refractory period, referred to as the plateau, is long. It lasts 100–300 ms, and the value of voltage is stable at approximately +15 mV. A balance between cations flowing in and out of the cell must be maintained during this phase. The flow of potassium cations out of the cell is counterbalanced by an inward flow of calcium cations. Thus, first, the heart muscle is protected against disablement of the pumping function via sustained (tetanic) muscular contraction. Second, this refractory period is longer than the duration of the spread of impulse over the whole heart; therefore, in a healthy individual, an impulse cannot return or spread in loops. Action potential on the membranes of the cells of the heart's conduction system does not exhibit the plateau phase. This is because depolarization of these cells is caused by the opening of calcium channels and a flow of calcium cations into the cell. Sodium cations are involved only minimally in depolarization.

As described above and as Fig. 3.3 implies, the extracellular space is relatively positive, whereas the intracellular space is negative. This is of major significance for the polarity of the pacing pulse. To evoke an above-threshold stimulus by pacing and depolarize the membranes, a negative pacing pulse has to be applied. As a result, the extracellular space, in which a pacing lead is placed, becomes more electronegative. Thus, the difference in potentials on the membrane is reduced, that is, it approaches a value of zero, which is above the threshold level.

3.7 Spread of Impulse

The ability to generate and spread an impulse is characteristic of some cardiac fibers that comprise the heart's conduction system. In contrast to the working myocardial fibers, they lack the ability to contract.

Impulses between cells are transmitted by local electric currents determined by the gradients between depolarized and polarized (de-excited) regions. An impulse is propagated along the myocardial fibers over the whole heart. Cells are connected to each other by intercalated discs with minimal electric resistance. Any impulse with an above-threshold intensity then spreads over the whole heart and produces depolarization in all cells.

In a healthy heart, an impulse originates in the SA node and is propagated by the working atrial myocardium. No conduction pathways (such as those in the ventricles) are morphologically apparent here; however, there are certain preferred ways impulses can be spread. The AV node, which slows the impulse to a propagation velocity of about 5 cm/s, is the only conductive link between the atria and the ventricles. The entirety of the ventricles is excited by an impulse that is propagated to the Purkinje fibers by passing through the interventricular septum, AV bundle of His, bundle branches, and subsequent branching.

In relation to the conduction system, one needs to be aware of the so-called cardiac automaticity gradient that can be encountered while using the pacing method; it also has to be taken into account in certain patients when, for example, ventricular pacing is used. It refers to the fact that not only the SA node, but each component of the above-described conduction system, has the capacity to automatically and independently generate impulses, although these impulses have a lower rate. If the rate of production of spontaneous impulses in the SA node is 60–80/min, then the production is 40–60/min in AV node impulses and only 20 40/min for those generated in the ventricular conduction system. Thus, under normal conditions, generation of impulses in the sinus node predominates because it is fastest. Lower parts of the conduction system are involved in impulse generation only when there is a pathological loss of function of a superior node or when a conduction disorder occurs.

3.8 Origin of an Electrocardiogram

During conventional extracellular sensing, the depolarized region is electronegative with respect to the polarized regions. The excited myocardial fibers behave as a dipole and create elementary electrical fields that are summarily characterized by a vector of the cardiac electrical field. Electric phenomena can be traced either by leads (epimyocardial or endocardial) implanted in the heart or superficially from the limbs and chest. When the junction of sensing surface electrodes is located in a direction approximately parallel to the longitudinal axis of the heart, an electrocardiogram (ECG) has a characteristic appearance, as shown in Fig. 3.4. For the sake of good reproducibility, the placement of surface electrodes is standardized to the well-established bipolar limb leads (Einthoven), unipolar augmented limb leads (Goldberg), and unipolar chest leads (Wilson). In pacing practice, which is particularly aimed at distinguishing paced and intrinsic cardiac activity, measuring the width of the QRS complex, or both, this standardization is not adhered to often. The electrical activity of the heart, defined by the sum of action potentials in all cells, can, of course, be recorded inside the heart. An intracardial recording of the electrical activity is referred to as an electrogram. It is obtained through implanted leads on the programmer screen.

A normal cardiac cycle begins with a small, rounded, positive P wave. It lasts about 80 ms and represents depolarization of the atria. The direction of the instantaneous vector of the electrical field is downward and to the left. It is followed by a PQ segment determined by the isoelectric line, and it also lasts about 80 ms. Next, the ventricular complex consisting of QRS waves and a T wave follows. A negative Q wave represents the onset of depolarization of the ventricular myocardium in the septal region. The direction of the instantaneous vector of the electrical field is downward and to the right. A prominent and positive R wave indicates propagation of an impulse over the walls of the ventricles; a negative S wave represents activation of the ventricular myocardium at the base of the left ventricle, and the instantaneous vector of the electrical field points to the left. This segment lasts

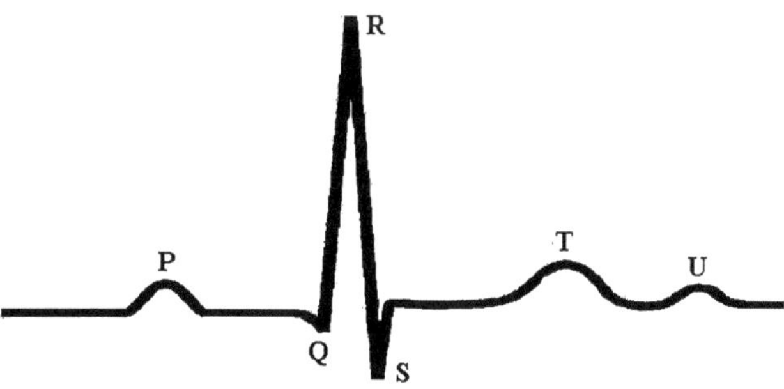

Fig. 3.4 Surface ECG, basic form

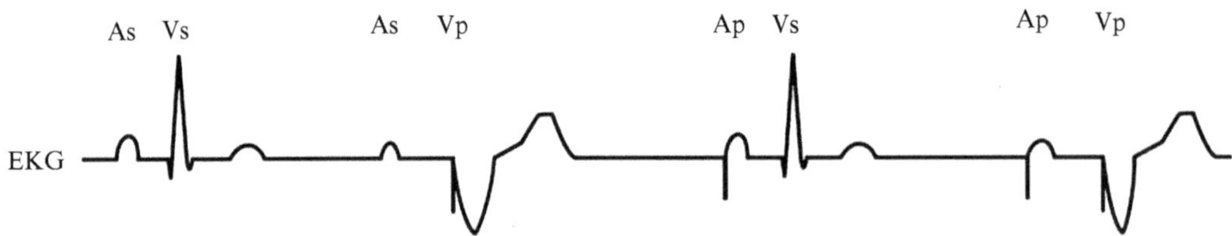

Fig. 3.5 Surface ECG at different paced and intrinsic heart beat combinations

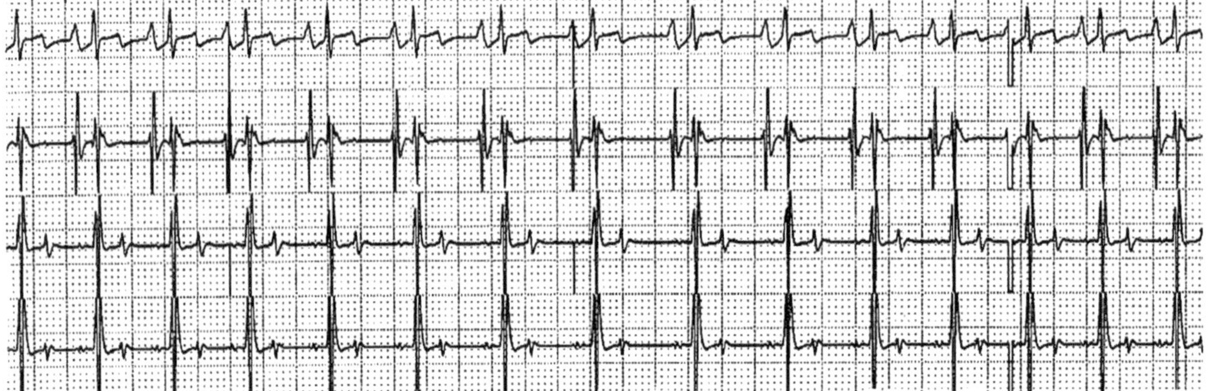

Fig. 3.6 Intracardial electrogram

about 100 ms and is followed by the ST segment, which is an isoelectric line lasting about 120 ms and corresponding to the plateau phase of the membrane potential. The final T wave lasts about 160 ms and represents repolarization of the ventricular myocardium proceeding in the opposite direction. The T wave is sometimes followed by a U wave, probably caused by repolarization of the Purkinje fibers.

With respect to the use of the pacing method, the vulnerable period of the ventricular myocardium needs to be discussed. It occurs before the end of the T wave. If an above-threshold impulse acts on the ventricular myocardium in this phase of repolarization, the development of ventricular fibrillation is nearly certain. During ECG evaluation, fusion or pseudofusion heart beats can also be encountered. They occur when intrinsic cardiac activity and pacing pulse occur at the same or nearly the same time. If intrinsic activity is delayed, the resulting QRS complex will resemble a paced beat, and vice versa – if intrinsic activity precedes the pacing pulse, the QRS complex will look like an intrinsic rhythm. If the pacing pulse has no effect on the intrinsic QRS complex or T wave, the beat is referred to as pseudofusion. An occurrence of fused beats can be mistaken for a system failure. However, as the sensing of intrinsic activity with sensing circuits can take place only during the late QRS complex, a pacing pulse can be delivered in the meantime [94].

In terms of the pacing method, what is commonly encountered is not a physiological shape of an ECG curve, but rather a shape of paced heart rhythm, intracardial electrograms, or both. Figure 3.5 shows typical ECG waveforms with various combinations of intrinsic and paced cardiac activity in the right atrium and ventricle. The shape of paced right ventricular activity is a markedly negative wave. From the left to the right, it shows intrinsic P and R waves (a physiological shape), an intrinsic P wave and pacing in the ventricle, pacing in the atrium and an intrinsic R wave, and pacing in both the atrium and the ventricle.

When implanted devices are followed up, it is often necessary to interpret electrograms sensed by those devices by means of electrodes of the implanted leads. The curves of these intracardial electrograms are rather different from those of the superficial ones because of their placement as well as different input filter settings and, in particular, a short distance between the sensing electrodes. Nevertheless, it is sufficient for distinguishing between paced and intrinsic cardiac activity. In some implantable defibrillators, it is possible to observe an electrogram sensed by defibrillation lead electrodes. There is a longer distance between them, which results in a recording that has superior informative value than in the case of sensing by means of bipolar pacing leads (Fig. 3.6).

4 Pharmacological Treatment of Cardiac Rhythm Disorders

To provide a comprehensive picture of pacing, this chapter presents an overview of cardiac rhythm disorders with clinical implications as well as basic information on pharmacological treatment options.

4.1 Cardiac Rhythm Disorders

Cardiac rhythm disorders are generally referred to as arrhythmias. They can result from abnormal impulse initiation, abnormal impulse conduction, or a combination of both (Fig 4.1). In addition to myocardial damage, a disorder can be caused by extracardiac factors (changes in mineral or hormone levels, alcohol intoxication, medication use, etc.).

Arrhythmias can be categorized according to heart rate as bradycardia (a heart rate below 60 beats/min) and tachycardia (three or more cardiac cycles above 100 beats/min, with electrical activity of the heart arising from the same site). Arrhythmias also can be categorized according to the anatomic site of origin as sinus, supraventricular, and ventricular arrhythmias [23, 25]. Sinus and supraventricular arrhythmias originate above the branching of the bundle of His; ventricular arrhythmias arise from the ventricular myocardium. In supraventricular arrhythmias, the shape of the QRS complex on an electrocardiogram (ECG) is generally of normal width, whereas in ventricular arrhythmias it is widened.

Arrhythmias can be of cardiac or noncardiac primary etiology. Cardiac causes primarily include organic myocardial damage (coronary artery disease, cardiomyopathy, hypertrophy, inflammation, fibrosis, arrhythmogenic right ventricular dysplasia, congenital developmental defects with disorders of impulse generation and conduction); hemodynamic causes (congenital developmental defects with impaired hemodynamics, acquired valvular heart disease, constrictive pericarditis, ventricular septal defect); disorders due to medical intervention (cardiac surgery, radiofrequency ablation); and cell membrane disease (long QT syndrome and others). Noncardiac etiologies include changes in the internal environment (hypoxia, anemia, ion imbalance); endocrine causes (thyroid disease, adrenal disease); vegetative nervous system status (carotid sinus syndrome, neurocardiogenic syncope); and others (pulmonary embolism, intoxication).

Symptoms differ depending on the type of arrhythmia. In bradycardia, they usually include fatigue, dizziness, vertigo, muscle weakness, breathlessness, or fainting. When normal generation of electric impulses is impaired, the heart is unable to increase its activity during exercise. When the heart rate is very slow or there are pauses in electric impulse generation of up to a few seconds, a short loss of consciousness may occur. However, similar symptoms can be caused by other diseases. In tachycardia, palpitation is the most frequent symptom. It is an acutely perceived rapid heartbeat. Tachycardia also present as bouts of regular palpitations with a sudden onset and end. When there is concomitant involvement of the coronary arteries, chest pain may occur.

The principle arrhythmogenic mechanisms [23] are altered automaticity (enhanced normal automaticity, abnormal automaticity), triggered activity (early or delayed after-depolarization), and reentry. Enhanced automaticity is due to accelerated depolarization in the cells of the sinoatrial node or other cells of the conduction system, resulting in sinus tachycardia or atrial tachycardia. In abnormal automaticity, impulses are generated in the cells of the conduction system outside the sinus node or even in myocardial cells. This results in ectopic atrial tachycardia, accelerated idioventricular rhythm, or ventricular tachycardia. Triggered activity is caused by an abnormal process of repolarization that leads to further depolarization. The subsequent impulse is thus triggered by the previous impulse.

Early after-depolarization occurs before the completion of the previous repolarization, giving rise to a form of ventricular tachycardia called Torsades de pointes. Delayed after-depolarization occurs after the completion of the previous repolarization because of excessive intracellular concentration of calcium cations. This results in ventricular extrasystoles or some atrial and ventricular tachycardia.

D. Korpas, *Implantable Cardiac Devices Technology*, DOI 10.1007/978-1-4614-6907-0_4,

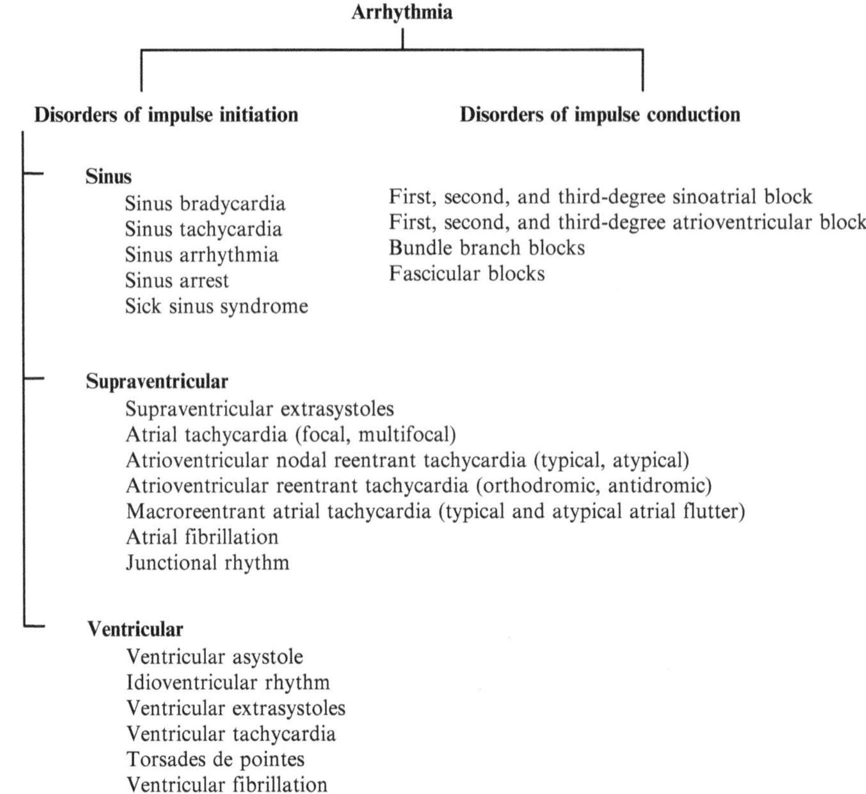

Fig. 4.1 Overview of arrhythmias

Another arrhythmogenic mechanism, and the most common, is reentry, that is, a reentrant (circulating) impulse. It may occur when two sites in the myocardium are connected by two pathways having different velocities of impulse propagation. This circulating impulse has a higher rate than the sinus node and is thus a predominant impulse generator. According to the size of the potential circuit, reentry is divided into macroreentry (large, occurring around the orifice of large vessels, the atrioventricular (AV) junction, or an ischemic focus) and microreentry (small, occurring in the area of Purkinje fibers). Reentry gives rise to all atrial and ventricular flutters, tachycardia, and fibrillation.

4.1.1 Sinus Arrhythmias

Sinus arrhythmias (Fig. 4.2) originate in the sinoatrial node [25].

4.1.1.1 Sinus Bradycardia

Sinus bradycardia is a heart rhythm that has a rate of under 60 beats/min. Physiologically, it occurs in trained athletes or during sleep. It can occur in cases of hypothermia, treatment with β-blockers, or intracranial hypertension. In symptomatic bradycardias, pacing therapy is the treatment of choice.

4.1.1.2 Sinus Tachycardia

Sinus tachycardia is a heart rhythm that has a rate of more than 100 beats/min. Physiologically, it occurs with physical or mental exertion. It may have a pharmacological cause. If it is more frequent, β-blockers are the treatment of choice.

4.1.1.3 Sinus Arrhythmia

Sinus arrhythmia is a markedly irregular heart rhythm. Physiologically, it occurs as respiratory sinus arrhythmia in which the heart rate accelerates during inspiration and slows during expiration.

4.1.1.4 Sinus Arrest

Sinus arrest is a transient complete loss of sinus node activity. Arrests lasting more than 3 s are considered pathological. When there are significant symptoms, pacing therapy is used for treatment.

4.1.1.5 Sick Sinus Syndrome

Sick sinus syndrome is a combination of sinus bradycardia and another arrhythmia, most commonly supraventricular tachycardia. It occurs in older age because of degenerative changes in the sinus node, more common ischemia and cardiac surgery, etc. It may even be caused pharmacologically. When a patient is symptomatic, pacing therapy is used for treatment.

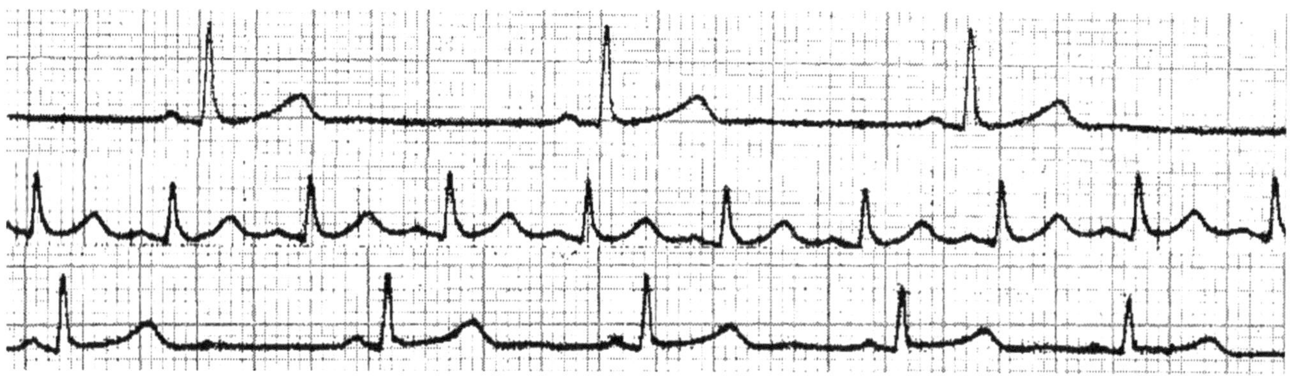

Fig. 4.2 Sinus arrhythmia [24] (Used with permission of C. Cihalik, Palacky University Olomouc, Czech Republic)

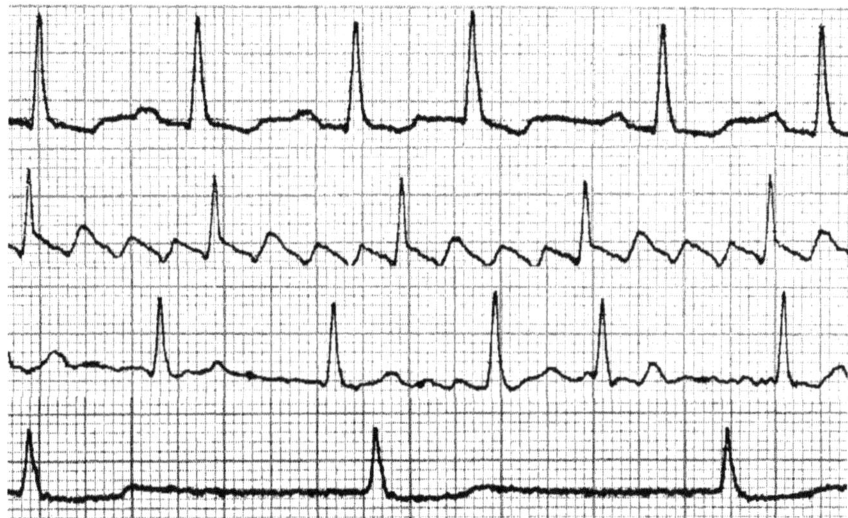

Fig. 4.3 Supraventricular arrhythmia [24] (Used with permission of C. Cihalik Palacky University Olomouc, Czech Republic)

4.1.2 Supraventricular Arrhythmias

Supraventricular arrhythmias (Fig. 4.3) originate and persist in the region of the atria, the sinoatrial and AV nodes, and accessory AV pathways [25].

4.1.2.1 Supraventricular Extrasystoles

Supraventricular extrasystoles are premature contractions originating from anywhere except the sinus node. Based on the source of pathological activity, extrasystoles can be divided into atrial, junctional, and ventricular. In supraventricular extrasystoles, the shape of the P wave depends on the distance from the sinus node. A supraventricular extrasystole (premature atrial contraction) is followed by an incomplete compensatory pause because the atrial impulse is usually conducted in a retrograde fashion to the sinus node and discharges the impulse being generated in it. A new impulse is then generated from the beginning and is further conducted in a normal interval.

4.1.2.2 Atrial Tachycardia

Atrial tachycardia originates from a focus anywhere in the atria except the sinus node. The most commonly involved mechanism is abnormal automaticity; less frequently, reentry is the mechanism. The heart rate may be more than 200 beats/min. The shape of the P wave depends on the distance from the sinus node. The condition is treated with antiarrhythmic agents.

4.1.2.3 Atrioventricular Nodal Reentrant Tachycardia

AV nodal reentrant tachycardia (AVNRT) is the most frequent regular paroxysmal tachycardia originating from the atrial part of the AV node. It is based on a reentry mechanism and has a sudden onset and end. The most common form has a slow conduction from the atrium to the ventricle and a rapid conduction back. Atrial and ventricular activation occur simultaneously and, on a surface ECG, the P waves are hidden in the QRS complex or follow it immediately.

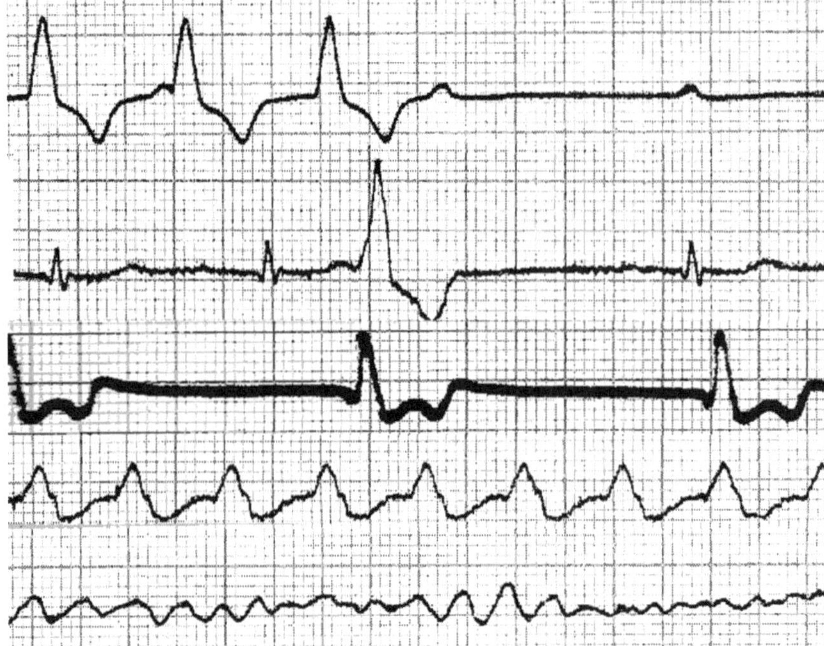

Fig. 4.4 Ventricular arrhythmia [24] (Used with permission of C. Cihalik Palacky University Olomouc, Czech Republic)

The rate of this tachycardia is 150–200 beats/min. Currently, the usual method of treatment of AVNRT is radiofrequency catheter ablation. An acute attack is terminated by vagal maneuvers or pharmacologically.

4.1.2.4 Atrioventricular Reentrant Tachycardia

AV reentrant tachycardia (AVRT) is caused by the presence of a pathway that actually short-circuits an electrically inert fibrous ring separating the atria from the ventricles. The ring is an insulating tissue and electrically separates the atria from the ventricles. The accessory pathway can cause the development of macroreentry by an impulse conducted from the atria to the ventricles by the AV node and returned to the atrium via the pathway, or vice versa. The conductive pathway can cause the development of another supraventricular tachycardia by conducting atrial activity to the ventricles, with a risk of developing ventricular fibrillation. On a surface ECG, the P waves follow the QRS complex. The rate of this tachycardia is 150–250 beats/min. The method of treatment of AVRT is radiofrequency catheter ablation. An acute attack can be terminated by vagal maneuvers or pharmacologically.

4.1.2.5 Atrial Flutter

Atrial flutter is a regular monomorphic tachycardia caused by a reentry mechanism. In its typical form, an impulse circulates in an counterclockwise direction in a defined tissue circuit in the right atrium. It may often co-occur with another type of arrhythmia or as a combination of atrial flutter and fibrillation. An ECG curve with negative saw-tooth flutter waves is characteristic. The rate of atrial flutter ranges from approximately 220 to 320 beats/min. Possible treatments include some antiarrhythmic agents, electrical cardioversion, and overdrive pacing with a rate higher than the flutter rate. Radiofrequency catheter ablation is also a method of choice.

4.1.2.6 Atrial Fibrillation

Atrial fibrillation is the most common clinically significant arrhythmia, the prevalence of which increases with age. It is also based on a reentry mechanism. There is a high risk of embolism. With a longer duration of atrial fibrillation, electrical and mechanical remodeling of the atrium occurs. No P waves are discernible on the ECG curve, and the isoelectric line is irregular. The rate of the fibrillation waves ranges from approximately 340 to 600/min, with an irregular conduction to the ventricles. Treatment with certain antiarrhythmic agents as well as electrical cardioversion is possible. When classic treatment methods fail, radiofrequency catheter ablation in the left atrium and pulmonary vein isolation are the methods of choice.

4.1.2.7 Junctional Rhythm

Junctional rhythm typically occurs only in sinus node abnormalities. Its rate is 35–50 beats/min. If faster, it is an active junctional rhythm or junctional tachycardia. In sustained forms, treatment is the same as that for bradycardias or AV blocks (i.e., cardiac pacing).

4.1.3 Ventricular Arrhythmias

Ventricular arrhythmias (Fig. 4.4) include a number of disorders, which often are very serious in terms prognosis. The underlying heart disease, left ventricular function, and the condition of the coronary bed always play an important role. For sustained ventricular tachycardia to occur, an initial abnormal stimulus and maintenance of the initiated arrhythmia are required. The most common mechanism of maintaining arrhythmia is reentry. Less frequently, it involves repeated initiation by automaticity or triggering.

4.1.3.1 Ventricular Asystole

The ventricular myocardium is not electrically activated. Only P waves or the isoelectric line are seen on the ECG. Transient arrests are characteristic of sick sinus syndrome.

4.1.3.2 Idioventricular Rhythm

Heart action is activated only by tertiary impulse generation from the ventricles. Heart rate decreases to 30–40 beats/min.

4.1.3.3 Ventricular Extrasystoles

Ventricular extrasystoles (premature ventricular contraction) arise in ectopic ventricular foci, most commonly in the ventricular musculature or Purkinje fibers. They are more serious than supraventricular extrasystoles but can occur in people with no organic heart disease. The incidence increases with age. Depending on the frequency of their occurrence, extrasystoles are categorized as single, multiple, and coupling. According to the number of ectopic foci, they are classified as unifocal (from one focus, a constantly identical shape on the ECG) and multifocal (from several foci, a varied shape). The QRS complex of ventricular extrasystoles is always widened, whereas in supraventricular extrasystoles it is mostly narrow. A ventricular extrasystole usually does not propagate back to the atria and propagation of sinus impulses is not impaired. A ventricular extrasystole is followed by a full compensatory pause. This is because the ventricles are still in the refractory period when the subsequent sinus impulse is propagated after an extrasystole and therefore only contract with the next sinus impulse. The sum of intervals before and after the extrasystole is equal to two normal cardiac cycles. Ventricular extrasystoles are most typically manifested by palpitations. The most severe form of ventricular extrasystoles is the so-called R-on-T phenomenon, in which an extrasystole in the vulnerable period can trigger ventricular fibrillation.

4.1.3.4 Ventricular Tachycardia

Ventricular tachycardia (VT) refers to a sequence of three to five or more successive ventricular extrasystoles. Pathological impulses activate the ventricles at a rate of 140–220/min. According to the morphology of the QRS complex on the ECG, VT is classified as monomorphic (identical QRS complexes) or polymorphic (varying QRS morphology). The width of the QRS complex is more than 0.12 s. Sustained (more than 30 s) and unsustained (three to ten cardiac intervals) VTs are distinguished according to duration. VT occurs as a late complication of acute myocardial infarction and coronary artery disease or as a response to overdose with certain medications. Polymorphic VT with underlying congenital or acquired ion channel disorders is a distinct group.

4.1.3.5 Torsades De Pointes

The shape of this polymorphic VT on the ECG is characteristic and resembles the resultant modulated signal of the amplitude modulation, which is due to a periodic change in the QRS vector around the isoelectric line in the frontal plane. The rate is 200 beats/min or more. Early after-depolarization is the likely arrhythmogenic mechanism. Torsades de pointes occurs in relation to a prolonged QT interval.

4.1.3.6 Ventricular Fibrillation

In ventricular fibrillation (VF), circulatory arrest occurs because the ventricular myocardium does not contract effectively. Uncoordinated electrical activity only results in hemodynamically ineffective contractions. VF results from an underlying heart muscle disorder with regions of inhomogeneous refractoriness. On the ECG, the individual waves are not discernible, except for the fibrillation waves, frequently with decaying amplitude.

4.1.4 Disorders of Impulse Conduction

Disorders of impulse conduction (Fig. 4.5) can occur at all levels of the conduction system [25]. Thus, sinoatrial, AV, bundle branch, and fascicular blocks are distinguished. Impulse conduction can also be interrupted at several levels. Typical ECG waveforms presented for the respective cardiac rhythm and conduction disorders will be of particular use in more complex applications of dual-chamber pacing.

4.1.4.1 Sinoatrial Block

In sinoatrial (SA) block, impulse conduction from the sinus node to the atria is impaired (slowed or blocked). As in the case of AV block (described below), there are three degrees of SA block. Because the sinus node signal is not apparent on a surface ECG, it is possible to diagnose only second-degree block when there is progressive shortening of the PP interval until one interval is lengthened and the QRS complex is dropped. Of more importance is third-degree block, when no impulse is conducted from the sinus node to the atria and only the isoelectric line is visible on the ECG, and, subsequently, junctional rhythm develops. SA block occurs in organic heart disease; some medications may have an adverse effect.

4.1.4.2 Atrioventricular Block

The region of the AV junction comprises the AV node (the proximal portion), the bundle of His, and the beginning of both bundle branches (the distal portion). Thus, in AV block, impulse conduction from the atria to the ventricles is impaired

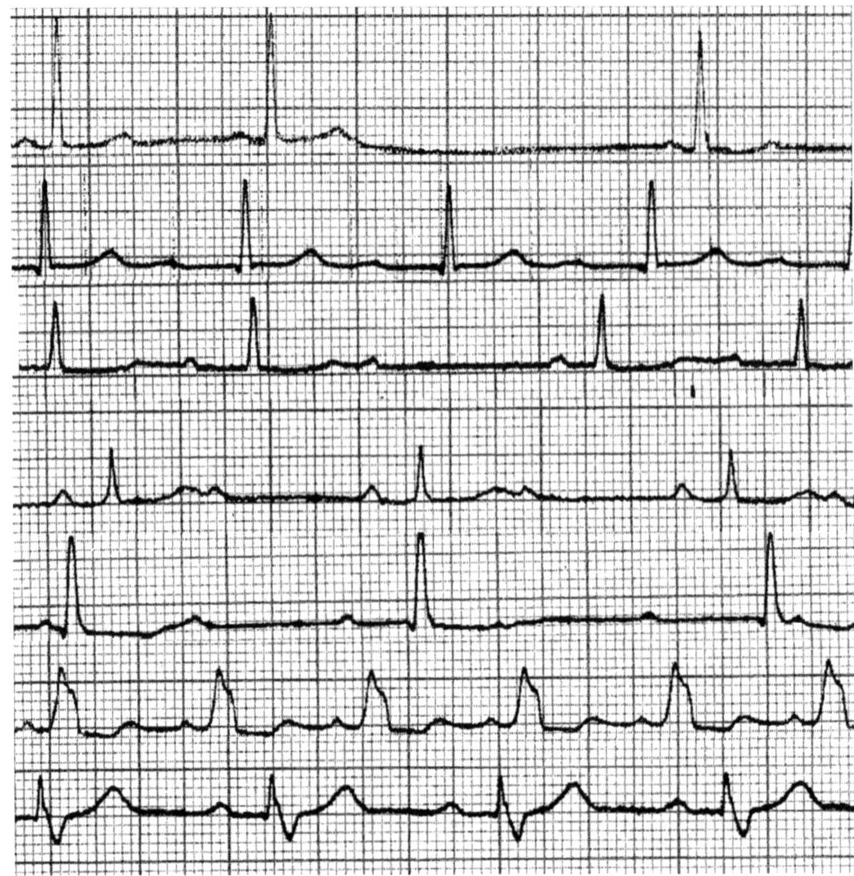

Fig. 4.5 Conduction disturbances [24] (Used with permission of C. Cihalik Palacky University Olomouc, Czech Republic)

(slowed or blocked). As in the case of SA block, described above, there are three degrees of AV block that can combine with each other. The particular degrees differ in the proportion of conducted and unconducted impulses. Third-degree AV block, in which no impulse is conducted from the atria to the ventricles, is the most serious form.

4.1.4.3 First-Degree AV Block

Impulse conduction from the atria to the ventricles is slowed in first-degree AV block, which is reflected on the ECG as a prolongation of the PQ interval more than 0.2 s, with the length of the PQ interval being constant. It usually ranges from 0.2 to 0.4 s, and in rare cases up to 1 s. Isolated first-degree AV block is not treated.

4.1.4.4 Type I Second-Degree AV Block (Mobitz I, Wenckebach)

In this type of block, there is progressive lengthening of the PQ interval, ultimately leading to a dropped QRS complex. The progressive prolongation of the PQ interval ultimately results in a P wave that is not followed by a QRS complex. After the dropped QRS complex, AV conduction recovers; thus the whole process is repeated regularly. The ratio of P waves to QRS complexes generally is n:(n – 1). The ECG difference between second-degree SA block and second-degree AV block is that, in second-degree SA block, the P wave as well as the QRS complex are dropped when there is a missed beat, whereas in second-degree AV block, only the QRS complex is dropped.

4.1.4.5 Type II Second-Degree AV Block (Mobitz II)

This block is characterized by a constant PQ interval and a subsequent sudden dropping of a QRS complex with a preserved P wave. The ratio of P waves to QRS complexes is generally n:1, or the QRS complex may be completely absent after several normal impulse conductions to the ventricles. It is a more serious block than Mobitz I because it may readily change into third-degree AV block.

4.1.4.6 Third-Degree AV Block

In third-degree AV block, conduction of impulses from the atria to the ventricles is blocked completely. No relationship exists between the atrial and ventricular activity. The atria are activated by the sinus node and the ventricles from a secondary focus. When the block is more proximal, it is a junctional rhythm with narrow QRS complexes; when the block is more distal, it presents as a slower idioventricular rhythm with wide QRS complexes.

4.1.4.7 Bundle Branch Blocks

Bundle branch blocks are the most frequent type of conduction disorders. They are caused by impaired conduction at the level of the right or left bundle branch and are present in approximately 0.6 % of the population, with a higher occurrence in the elderly. In bundle branch block, there is delayed activation of the myocardium of the corresponding ventricle. This delay leads to subsequent characteristic morphological changes in the QRS complex. Right bundle branch block (RBBB) commonly occurs in right ventricular disease and left bundle branch block in left ventricular dilation and hypertrophy.

4.1.4.8 Fascicular Blocks

Impulse conduction in the left ventricular myocardium occurs via two fascicles, anterior and posterior, that are continuations of the left bundle branch. The right bundle branch in the right ventricle is the third fascicle. Left anterior and left posterior fascicular blocks are distinguished. Bifascicular blocks are caused by a RBBB with concomitant left anterior or left posterior fascicular block. Trifascicular block is a combination of a RBBB and a block of the two fascicles of the left bundle branch.

4.2 Pharmacological Treatment of Arrhythmias

Despite the advent of nonpharmacological treatment that involves pacemaker implantation in bradycardia (e.g., radiofrequency catheter ablation in supraventricular tachycardia) and ICD implantation in ventricular arrhythmias [26, 27], pharmacotherapy still has a firm place in the long-term management of arrhythmias. However, the primary indications for pharmacotherapy are acute management of tachycardia and prevention of tachycardia in patients in whom ablation has failed, followed by atrial fibrillation and VT, for which pharmacotherapy is a complementary treatment to reduce the ICD shock rate. In practice, it is necessary to keep in mind some proarrhythmogenic effects of certain antiarrhythmic drugs.

4.2.1 Classification of Antiarrhythmic Drugs

Antiarrhythmic drugs affect certain ion channels responsible for the membrane potential of a cell. A change in the membrane potential affects depolarization and repolarization of the cell membrane and may thus affect the basic mechanisms of origin and maintenance of tachycardia. The slowing of depolarization avoids spontaneous depolarization, and slowing the rate of depolarization and repolarization avoids the maintenance of tachycardia in a reentry circuit. The Vaughan-Williams classification (Table 4.1) divides antiarrhythmic drugs into four classes based on their effect on a specific ion channel [23].

A limitation of this classification of antiarrhythmic drugs is that it does not include some drugs (digoxin, adenosine, magnesium sulfate). A more recent classification, called the "Sicilian gambit" [23], divides all known antiarrhythmic drugs according to the site where their effect is exerted (membrane channels, receptors, and pumps).

In practice, it is suitable to classify antiarrhythmic drugs according to the cardiac structure that is to be affected. Antiarrhythmic drugs are divided into two groups consisting of four drugs. The first includes AV nodal blockers that slow impulse conduction in the AV node. The second includes stabilizers that stabilize electrical processes in the atria and the ventricles (Table 4.2).

4.2.2 Pharmacological Treatment

Permanent treatment of bradycardias mainly involves nonpharmacological treatment using permanent pacing. To ascertain the method of treatment, it is necessary to determine the degree of symptomaticity and severity of bradycardia. Treatment is indicated only in symptomatic bradycardias [29]. The following conditions are considered symptomatic: those leading to cardiac arrest, syncopes with underlying SA or AV blocks, presyncopes, or collapses. The treatment of acute bradycardias involves administration of atropine or isoprenaline, application of temporary pacing, or both.

Table 4.1 The Vaughan-Williams classification of antiarrhythmic drugs [23, 28]

Class	Blocks	Effect on depolarization	Effect on repolarization	Drugs
Ia	Na^+ channel	Moderate prolonged	Prolonged	Quinidine, disopyramide, procainamide
Ib	Na^+ channel	Weak	Shortened	Mexiletine, lidocaine
Ic	Na^+ channel	Strong prolonged	Weak	Propafenone, flecainide
II	β receptors	Weak	Weak	Metoprolol, atenolol
III	K^+ channel	Weak		Amiodarone, bretylium, sotalol
IV	Ca^{2+} channel	Weak	Weak	Diltiazem, verapamil

Table 4.2 Classification according to interaction site

Node blockers	Stabilizers
Adenosine	Trimecaine
Calcium channel antagonists	Propafenone
β-Blockers	Sotalol
Digoxin	Amiodarone

According to guidelines [34], in a simplified fashion, pacemaker implantation is indicated for an asystole longer than 3 s, detection of a heart rate below 40 beats/min, and second- or third-degree AV block. When bradycardias are not clearly indicated for permanent pacing, no drugs are usually administered.

Pharmacological treatment of tachycardia involves administration of antiarrhythmic drugs [29, 30]. The original drugs were antiarrhythmic drugs of the Ia group. This group has numerous side effects and is not very efficacious, and some of the agents are no longer commercially available. Antiarrhythmic drugs of the Ic and III groups are better tolerated and more efficacious; however, they reduce myocardial contractility and affect the conductivity of the conduction system. In supraventricular tachycardia, such as AVRT, AVNRT, sinus tachycardia, atrial tachycardia, and atrial flutter, AV nodal blockers or propafenone are administered. In atrial fibrillation, AV nodal blockers (digoxin, verapamil, β-blockers) are used to control ventricular rate, and stabilizers (propafenone, sotalol, amiodarone, or, more recently dronedarone, a noniodinated amiodarone derivative) are used to control rhythm. Monomorphic VT with a wide QRS complex is treated with stabilizers (sotalol, amiodarone) over the long term. Amiodarone is the most commonly used agent in the long-term treatment of polymorphic VT. Table 4.3 presents an overview of pharmacological antiarrhythmic therapy. However, it is an adjunct treatment to ICD implantation to reduce the rate of shock.

Table 4.3 Overview of arrhythmia treatments

Arrhythmia	Antiarrhythmic treatment
Bradycardia	Atropine, isoprenaline
Atrial fibrillation/flutter	Propafenone, amiodarone, dronedarone, sotalol, digoxin, β-blockers, calcium channel blockers
Paroxysmal supraventricular tachycardia	Adenosine, calcium channel blockers, β-blockers, propafenone
Pre-excitation	Procainamide, ajmaline
Ventricular tachycardia	Lidocaine, amiodarone, β-blockers
Ventricular fibrillation	Adrenalin, amiodarone

5 Pacing Modes

Pacing mode offers a basic overview of the therapeutic possibilities of certain types of devices. On the basis of the number of physically attached leads, single-chamber and dual-chamber modes are determined. The programmability of the device also allows dual-chamber devices to be operated as single-chamber devices (although the reverse is, of course, impossible).

This chapter information about programmable modes available with pacemakers. Explanation of North American Society of Pacing and Electrophysiology (NASPE)/British Pacing and Electrophysiology Group codes used in pacemakers is shown in Table 5.1. At positions I and II there are identification of paced and sensed heart chambers. At position III there is the mode of response to sensed intrinsic heart action (P wave, R wave, or both). Char I (inhibited) means that when intrinsic heart activity is sensed in the heart chamber, no pacing pulse will be generated during the actual cardiac cycle. Char T (triggered) means that a pacing pulse is generated in reaction to intrinsic heart activity. At position IV there is identification of rhythm modulation (adaptive-rate pacing). The fifth position indicates whether multisite pacing is possible. Multisite pacing is defined as using of more than one site in a given heart chamber for pacing. For explanation of other North American Society of Pacing and Electrophysiology/British Pacing and Electrophysiology Group codes, see Bernstein et al. [31]. Figure 5.1 may help clinicians program the most appropriate pacing mode for a specific patient.

The possibility of programing pacing mode is limited according to the particular pacemaker model. Within VVI(R), AAI(R), SSI(R), V00(R), A00(R), and S00(R), adaptive-rate pacing is combined with single-chamber pacing. Within DDD(R), DDI(R), and D00(R) modes, adaptive-rate pacing is combined with dual-chamber pacing. Other available pacing modes are DDD, DDI, D00, VDD, VVI, AAI, SSI, V00, A00, S00, VVT, AAT, and SST. The 0D0, 0S0, and 000 modes generally are programmable as temporary modes and are used for diagnostic purposes [32].

5.1 A00 Mode

Pacing pulses are delivered asynchronously to the atrium at the lower rate limit (LRL) (Fig. 5.2). Intrinsic heart activity (P wave) does not inhibit or trigger pacing in the atrium. The A00 mode is either directly programmable or operative with the use of a magnet in AAI and AAT modes. The asynchronous A00 mode may be used intraoperatively to reduce the likelihood of inhibition during electrocautery [32].

5.2 AAI(R) Mode

In the AAI(R) mode, sensing and pacing occur only in the atrium (Fig. 5.3). In the absence of sensed events, pacing pulses will be delivered to the atrium at the programmed LRL (AAI) or at a rate indicated by the sensor (AAI[R]). A sensed P wave or a paced atrium event causes the pacemaker's escape interval to reset [32].

5.3 AAT Mode

In the AAT mode, pacing pulses will be delivered to the atrium at the LRL in the absence of sensed events (Fig. 5.4). Sensed events trigger the atrial pulse and resets the pacemaker escape interval. Using the AAT mode outside of a diagnostic setting is not recommended because of the potential for triggered pacing in response to oversensing [32].

5.4 V00(R) Mode

Pacing pulses will be delivered asynchronously to the ventricle at the LRL (V00) or at the sensor-indicated rate in the V00(R) mode. Spontaneous ventricular events (R wave) do not inhibit or trigger pacing in the ventricle. The V00

D. Korpas, *Implantable Cardiac Devices Technology*, DOI 10.1007/978-1-4614-6907-0_5,

Table 5.1 NASPE/BPEG generic (NBG) pacemaker codes [31]

Position	I	II	III	IV	V
Category	Chambers paced	Chambers sensed	Response to sensing	Pacing rate modulation	Multisite pacing
Letters	0 – None	0 – None	0 – None	R – Pacing rate modulation	0 – None
	A – Atrium	A – Atrium	T – Triggered		A – Atrium
	V – Ventricle	V – Ventricle	I – Inhibited		V – Ventricle
	D – Dual (A+V)	D – Dual (A+V)	D – Dual (T+I)		D – Dual (A+V)
Manufacturer specific	S – Single (A or V)	S – Single (A or V)			

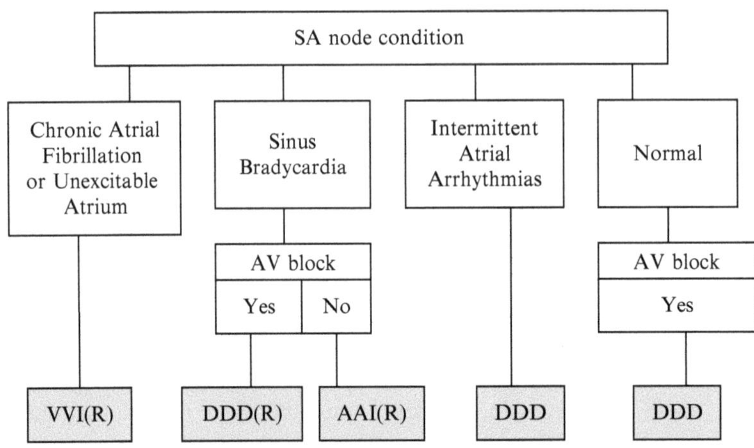

Fig. 5.1 Reference pacing mode decision tree [32] (© 2012 Boston Scientific Corporation or its affiliates. All rights reserved. Used with permission of Boston Scientific Corporation)

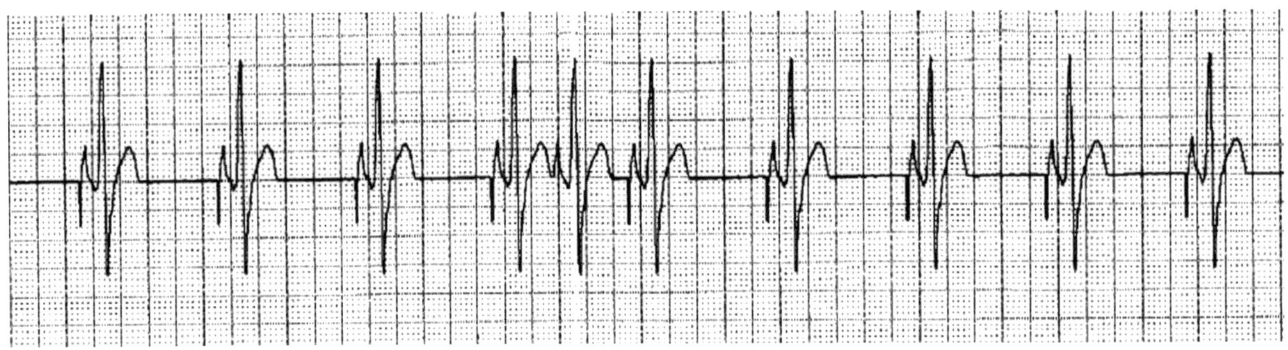

Fig. 5.2 A00(R) mode [32] (© 2012 Boston Scientific Corporation or its affiliates. All rights reserved. Used with permission of Boston Scientific Corporation)

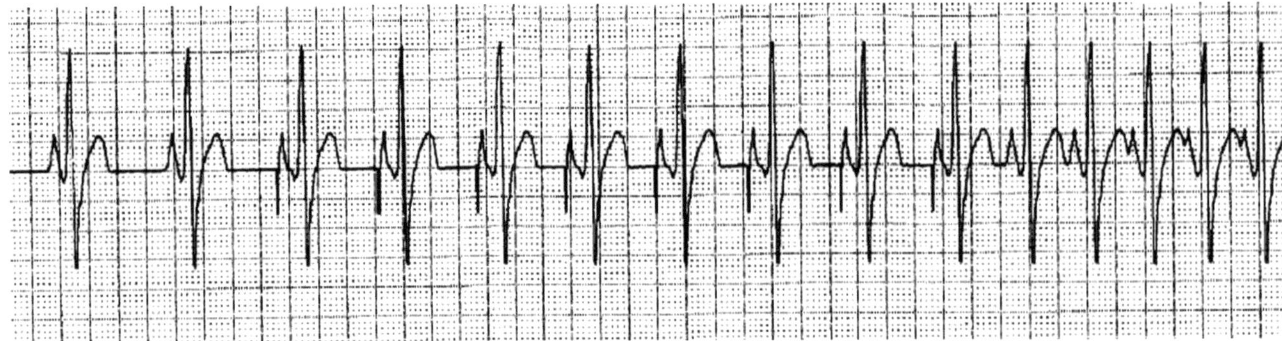

Fig. 5.3 AAI(R) mode [32] (© 2012 Boston Scientific Corporation or its affiliates. All rights reserved. Used with permission of Boston Scientific Corporation)

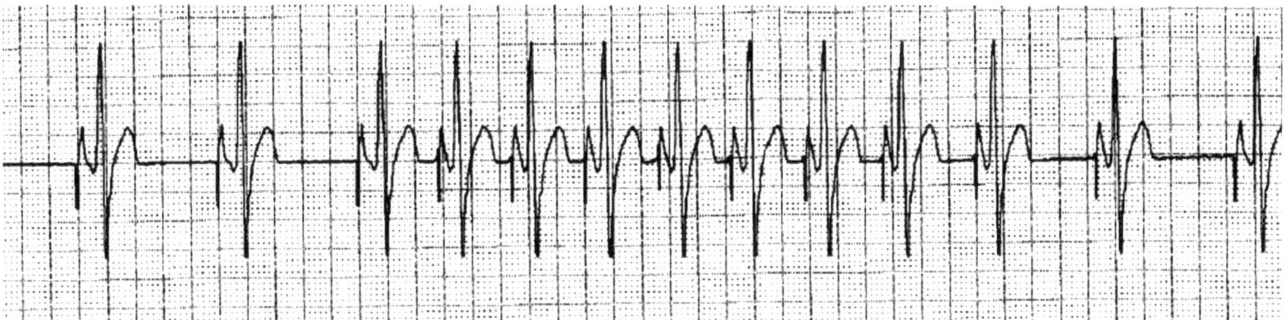

Fig. 5.4 AAT mode [32] (© 2012 Boston Scientific Corporation or its affiliates. All rights reserved. Used with permission of Boston Scientific Corporation)

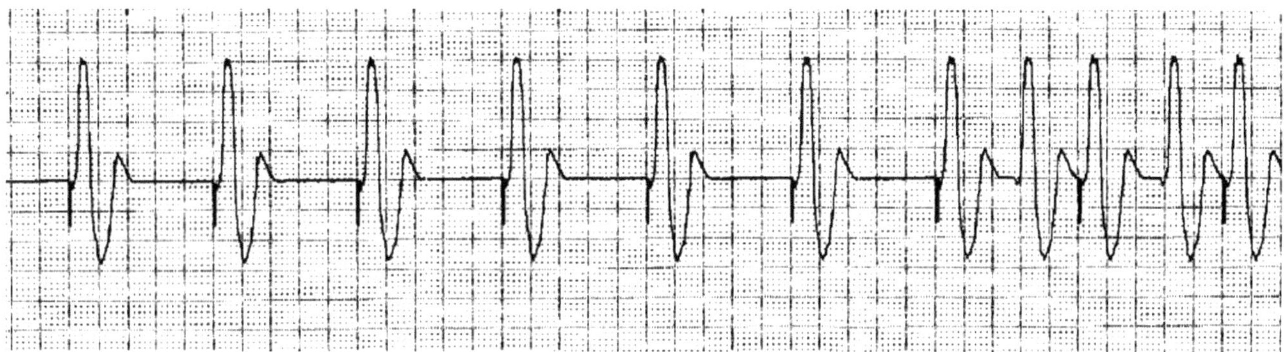

Fig. 5.5 V00(R) mode [32] (© 2012 Boston Scientific Corporation or its affiliates. All rights reserved. Used with permission of Boston Scientific Corporation)

mode is directly programmable and it is the magnet mode of VVI(R), V00(R), and VVT modes (Fig. 5.5). An asynchronous V00 mode may be used intraoperatively to reduce the likelihood of inhibition during electrocautery [32].

5.5 VVI(R) Mode

In the VVI(R) mode, sensing and pacing occur only in the ventricle (Fig. 5.6). In the absence of sensed events, pacing pulses will be delivered to the ventricle at the LRL (VVI) or at the sensor-indicated rate (VVI[R]). A sensed R wave or a paced ventricular event will reset the pacemaker's escape interval [32].

5.6 VVT Mode

In the absence of sensed events, pacing pulses will be delivered to the ventricle at the LRL in the VVT mode (Fig. 5.7). Sensed events trigger a ventricular pulse and reset the escape interval of the pacemaker. Using the VVT mode outside of a diagnostic setting is not recommended because of the potential for triggered pacing in response to oversensing [32].

5.7 VDD Mode

In the VDD mode, in the absence of sensed P or R waves, pacing pulses will be delivered to the ventricle at the programmed LRL (Fig. 5.8). A sensed P wave starts the atrioventricular (AV) delay interval. At the end of the AV delay interval, a ventricular pace will be delivered unless it is inhibited by a sensed R wave. A sensed R wave or a paced ventricular event will reset the pacemaker's escape interval [32]. This mode is a tracking mode.

5.8 VAT Mode

The VAT mode is the simplest (semiasynchronous) tracking mode. A sensed P wave starts the AV delay interval. At the end of the AV delay interval, a ventricular pace will be delivered (Fig. 5.9). However, the ventricle sensing is not active, and therefore there is a risk of vulnerable phase pacing in the ventricle.

5.9 D00(R) Mode

Pacing pulses will be delivered asynchronously to the atrium and the ventricle at the LRL (D00) or the sensor-indicated rate (D00[R]); pulses are separated by the programmed AV

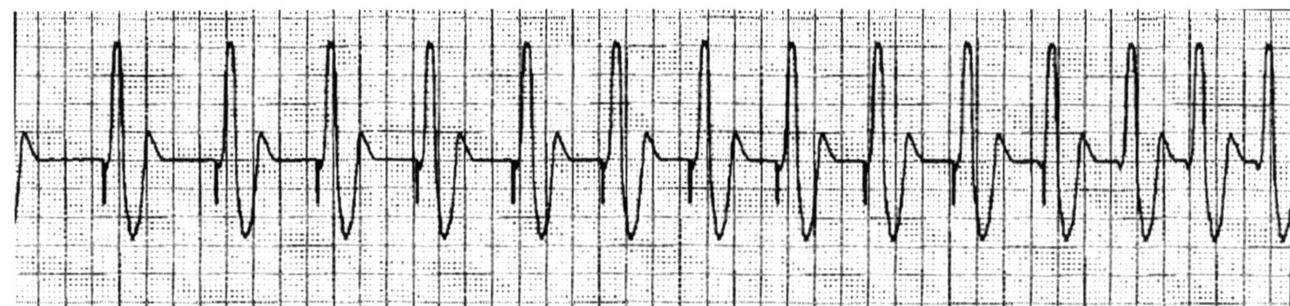

Fig. 5.6 VVI(R) mode [32]

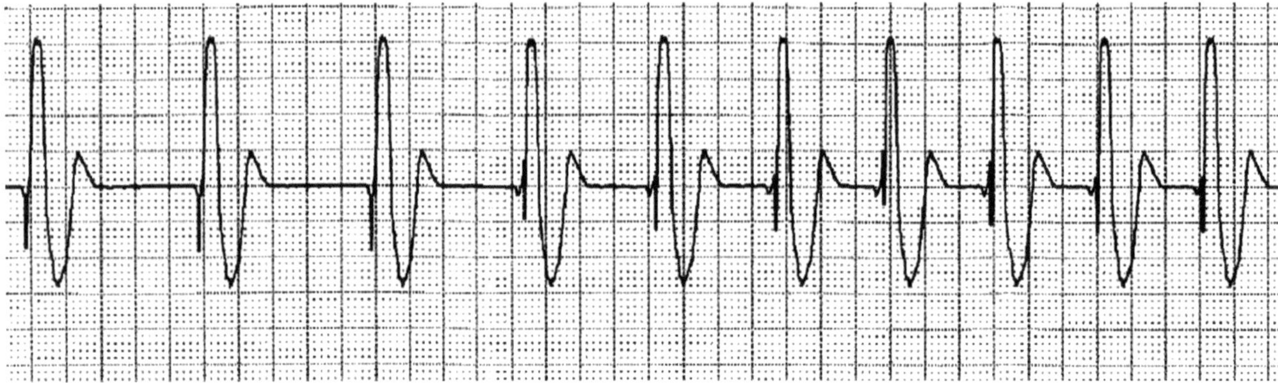

Fig. 5.7 VVT mode [32]

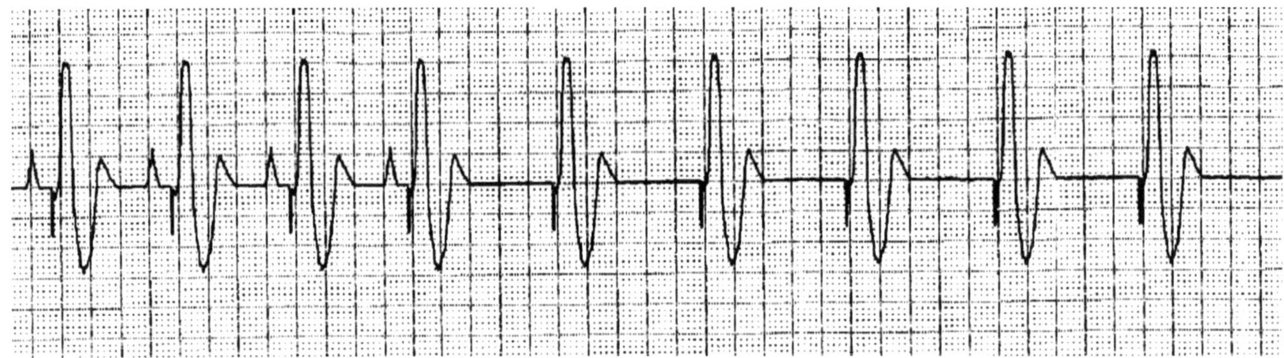

Fig. 5.8 VDD mode [32]

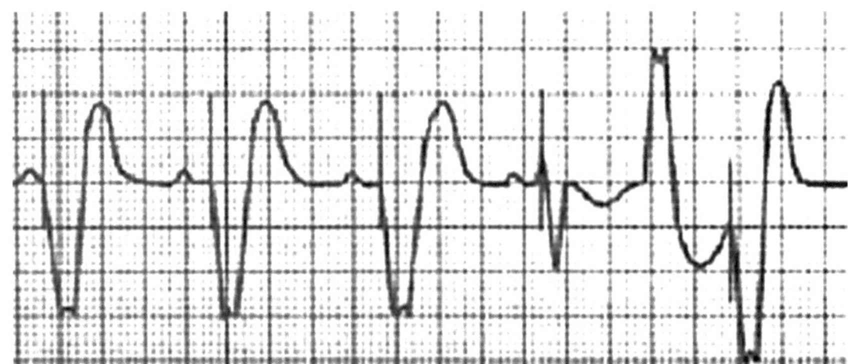

Fig. 5.9 VAT mode [32]

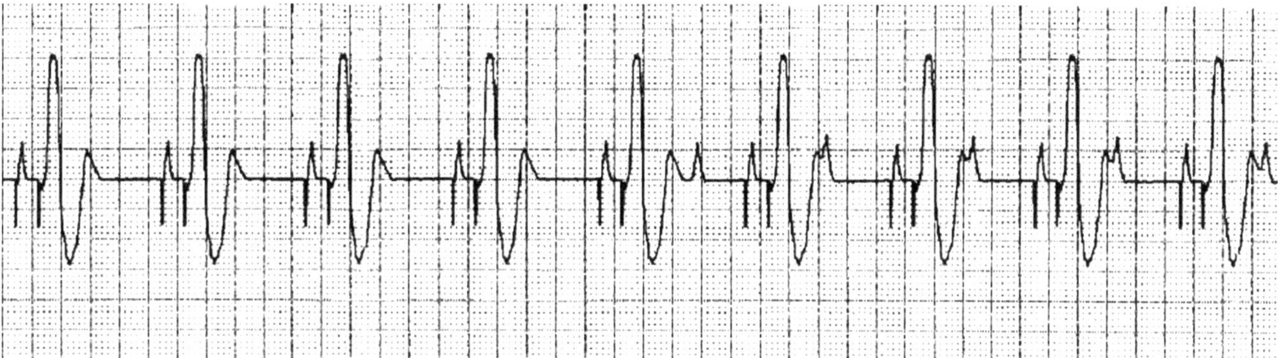

Fig. 5.10 D00(R) mode [32] ()

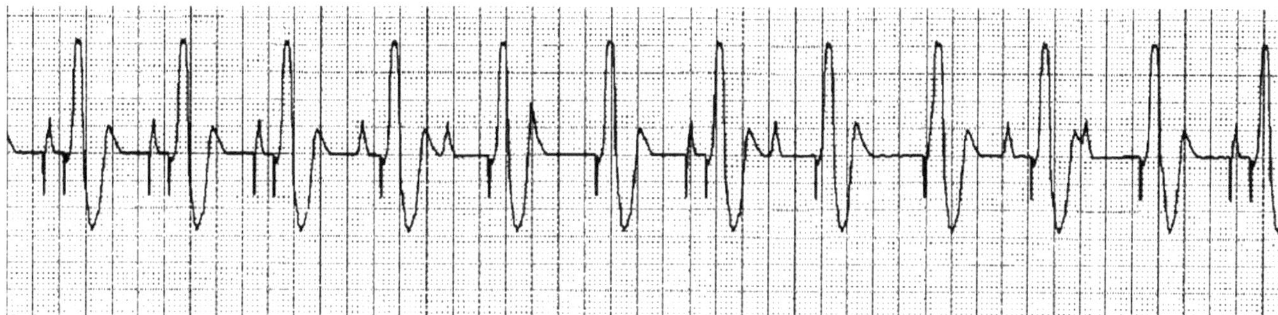

Fig. 5.11 DDI(R) mode [32] ()

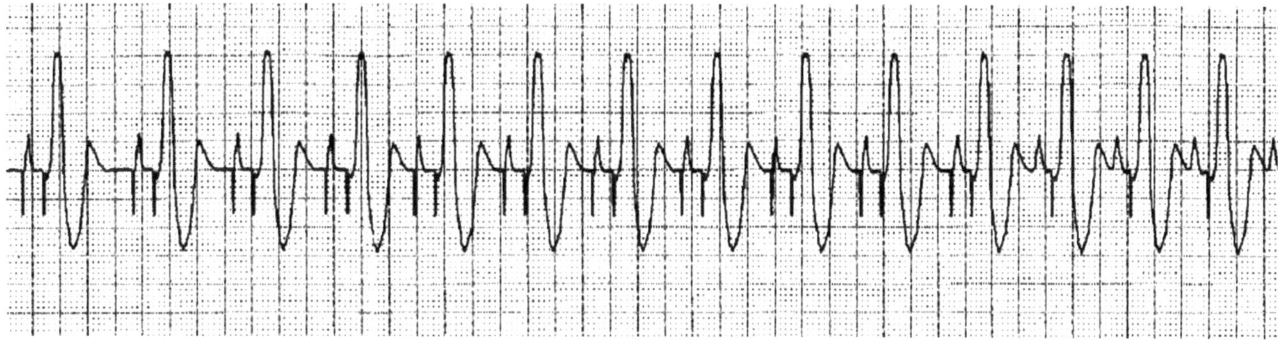

Fig. 5.12 DDD(R) mode [32] ()

delay (Fig. 5.10). Intrinsic events would neither inhibit nor trigger pacing in either chamber. The D00 mode is directly programmable and is the magnet mode of the corresponding dual-chamber modes, except for V00, which is the magnet mode for the VDD mode. The D00 mode may be used intraoperatively to reduce the likelihood of inhibition during electrocautery [32]. This mode is not a tracking mode.

5.10 DDI(R) Mode

In the absence of sensed P and R waves, pacing pulses will be delivered to the atrium and the ventricle at the LRL (DDI) or at the sensor-indicated rate (DDI[R]); pulses are separated by the programmed AV delay interval (Fig. 5.11). A sensed P wave will inhibit the atrial pace but will not start the AV delay [32]. Therefore, this mode is not a tracking mode.

5.11 DDD(R) Mode

In the absence of sensed P and R waves, pacing pulses will be delivered to the atrium and the ventricle at the LRL (DDD) or at the sensor-indicated rate (DDD[R]); pulses are separated by the programmed AV delay interval (Fig. 5.12). A sensed P wave will inhibit an atrial pace and start the AV delay. At the end of the AV delay, a ventricular pace will be delivered unless it is inhibited by a sensed R wave [32]. This mode is a tracking mode.

6 Indications for Implantable System Treatment

Implantation of pacing or defibrillation systems is a serious medical intervention. From this reason, patients have to fulfill certain medical criteria – called *indications* – before they undergo an implant. Indications are a complex of diagnostic and clinical symptoms presenting within the patient and indicate the suitability of an implantable system according to recent scientific knowledge. According to the indications it is determined which implantable system the patient needs; health insurance companies approve payment for implantation based on the fulfillment of indication criteria.

Before each implant, risks and benefits for the given patient are evaluated. The main criteria are life expectancy, quality of life, and patient prognosis. Recent recommendations use standard recommendation categorizations according to classes (Table 6.1) and three levels of evidence (Table 6.2) [33]. This classification enables easier decision making by the clinician.

Indication criteria can be generally divided into two categories: primary preventive (prevention before an event) and secondary preventive (prevention after an event). This division is most often associated with indications for use of implantable cardioverter-defibrillator (ICD) systems. The means of primary and secondary prevention are frequently the same.

The aim of primary preventive indications is to identify high-risk patients and implant an appropriate system even in them before first arrhythmic incident. Clinical trials have evaluated the risks and benefits of the ICD in the prevention of sudden cardiac death (SCD) and results have shown improved survival in multiple patient populations. Prospective registry data offer recommendations for ICD implantation in select other patient populations, such as those with hypertrophic cardiomyopathy, arrhythmogenic right ventricular dysplasia/cardiomyopathy, and long QT syndrome. In less common conditions (e.g., Brugada syndrome, catecholaminergic polymorphic ventricular tachycardia (VT), cardiac sarcoidosis, left ventricular (LV) noncompaction), clinical reports and retrospectively analyzed series provide less rigorous evidence in support of current recommendations for ICD use, but this constitutes the best available evidence for these conditions [34, 35]. For specification of arrhythmia classification, electrophysiology study is used. Its aim is to induce VT and determine its characteristics. The positive electrophysiology study stands for increased risk of future malignant arrhythmia. In addition, other risk factors such as family history or age are applied in primary prevention.

Secondary prevention refers to prevention of SCD in those patients who have already survived a prior sudden cardiac arrest or sustained arrhythmic episode, including hemodynamically unstable monomorphic VT, polymorphic VT, or ventricular fibrillation. The aim is to prevent the repetition of a risky situation.

6.1 Pacemaker Implantation Indications

Pacemakers generally are indicated for treatment of the following clinical conditions [32]:

- Symptomatic paroxysmal or permanent second- or third-degree atrioventricular (AV) block
- Symptomatic bilateral bundle branch block
- Symptomatic paroxysmal or transient sinus node dysfunction with or without associated AV conduction disorders (e.g., sinus bradycardia, sinus arrest, sinoatrial block)
- Bradycardia-tachycardia syndrome and to prevent symptomatic bradycardia or some forms of symptomatic tachycardia
- Neurovascular (vasovagal) syndromes or hypersensitive carotid sinus syndromes

Adaptive-rate pacing is indicated for patients who may benefit from increased pacing rates concurrent with increases in minute ventilation (MV), level of physical activity, or both. Dual-chamber and atrial tracking modes also are indicated for patients who may benefit from maintenance of AV synchrony. Dual-chamber modes are indicated specifically for treatment of the following:

D. Korpas, *Implantable Cardiac Devices Technology*, DOI 10.1007/978-1-4614-6907-0_6,

Table 6.1 Recommendation classifications

Class I	Conditions for which there is evidence, general agreement, or both that a given procedure or treatment is useful and effective
Class II	Conditions for which there is conflicting evidence, a divergence of opinion, or both about the usefulness/efficacy of a procedure or treatment
Class IIa	Weight of evidence/opinion is in favor of usefulness/efficacy
Class IIb	Usefulness/efficacy less well-established by evidence/opinion
Class III	Conditions for which there is evidence, general agreement, or both that the procedure/treatment is not useful/effective and in some cases may be harmful

Table 6.2 Levels of evidence

Level of evidence	Basis of recommendations
A	Evidence from multiple randomized trials or meta-analyses
B	Evidence from a single randomized trial or nonrandomized studies
C	Expert opinion, case studies, or standards of care

- Conduction disorders that require restoration of AV synchrony, including varying degrees of AV block
- VVI intolerance (e.g., pacemaker syndrome) in the presence of a persistent sinus rhythm
- Low cardiac output or congestive heart failure secondary to bradycardia

According to actual American College of Cardiology (ACC)/American Heart Association (AHA)/Heart Rhythm Society (HRS) 2008 Guidelines for Device-Based Therapy of Cardiac Rhythm Abnormalities [34, 35], the indications (class I and class IIa) for permanent pacing are described in the following paragraphs.

Recommendations for Permanent Pacing in Sinus Node Dysfunction

- Sinus node dysfunction (SND) with documented symptomatic bradycardia, including frequent sinus pauses that produce symptoms. For some patients this is the result of necessary long-term pharmaceutical therapy without the possibility of a dose change or use of an alternative therapy.
- Symptomatic sinus bradycardia that results from required drug therapy for medical conditions.
- Symptomatic chronotropic incompetence (inability to increase the heart rate with increased physical activity or other demand of the patient).
- SND with heart rate less than 40 beats/min when a clear association between significant symptoms consistent with bradycardia and the actual presence of bradycardia has not been documented.
- Syncope of unexplained origin when clinically significant abnormalities of sinus node function are discovered or provoked in electrophysiologic testing.

Recommendations for Acquired Atrioventricular Block in Adults

- Third-degree and advanced second-degree AV block at any anatomic level associated with bradycardia with symptoms (including heart failure) or ventricular arrhythmias presumed to be due to AV block
- Third-degree and advanced second-degree AV block at any anatomic level associated with arrhythmias and other medical conditions that require drug therapy that results in symptomatic bradycardia
- Third-degree and advanced second-degree AV block at any anatomic level in awake, symptom-free patients in sinus rhythm, with documented periods of asystole greater than or equal to 3.0 s, or any escape rate less than 40 beats/min, or an escape rhythm that is below the AV node
- Third-degree and advanced second-degree AV block at any anatomic level in awake, symptom-free patients with atrial fibrillation and bradycardia with one or more pauses of at least 5 s or longer
- Third-degree and advanced second-degree AV block at any anatomic level after catheter ablation of the AV junction
- Third-degree and advanced second-degree AV block at any anatomic level associated with postoperative AV block that is not expected to resolve after cardiac surgery
- Third-degree and advanced second-degree AV block at any anatomic level associated with neuromuscular diseases with AV block, such as myotonic muscular dystrophy, Kearns-Sayre syndrome, Erb dystrophy (limb-girdle muscular dystrophy), and peroneal muscular atrophy, with or without symptoms
- Second-degree AV block with associated symptomatic bradycardia, regardless of type or site of block
- Asymptomatic, persistent third-degree AV block at any anatomic site with average awake ventricular rates of 40 beats/min or faster if cardiomegaly or LV dysfunction is present or if the site of block is below the AV node
- Second- or third-degree AV block during exercise in the absence of myocardial ischemia
- Persistent third-degree AV block with an escape rate greater than 40 beats/min in asymptomatic adult patients without cardiomegaly
- Asymptomatic second-degree AV block at intra- or infra-His levels found during electrophysiologic testing
- First- or second-degree AV block with symptoms similar to those of pacemaker syndrome or hemodynamic compromise
- Asymptomatic type II second-degree AV block with a narrow QRS

Recommendations for Permanent Pacing After the Acute Phase of Myocardial Infarction

- Persistent second-degree AV block in the His–Purkinje system with alternating bundle branch block or third-degree

AV block within or below the His–Purkinje system after ST-segment elevation myocardial infarction
- Transient, advanced second- or third-degree infranodal AV block and associated bundle branch block. If the site of block is uncertain, electrophysiologic testing may be necessary
- Persistent and symptomatic second- or third-degree AV block

Recommendations for Permanent Pacing in Chronic Bifascicular Block
- Advanced second-degree AV block or intermittent third-degree AV block
- Type II second-degree AV block
- Alternating bundle branch block
- Syncope not demonstrated to be due to AV block when other likely causes have been excluded, specifically VT
- Incidental finding on electrophysiologic testing of a markedly prolonged HV interval (≥100 ms) in asymptomatic patients
- Incidental finding on electrophysiological study of pacing-induced infra-His block that is not physiological

Recommendations for Permanent Pacing in Hypersensitive Carotid Sinus Syndrome and Neurocardiogenic Syncope
- Recurrent syncope caused by spontaneously occurring carotid sinus stimulation and carotid sinus pressure that induces ventricular asystole of more than 3 s
- Syncope without clear, provocative events and with a hypersensitive cardioinhibitory response of 3 s or longer

Recommendations for Permanent Pacing in Children, Adolescents, and Patients with Congenital Heart Disease
- Advanced second- or third-degree AV block associated with symptomatic bradycardia, ventricular dysfunction, or low cardiac output
- SND with correlation of symptoms during age-inappropriate bradycardia; the definition of bradycardia varies with the patient's age and expected heart rate
- Postoperative, advanced second- or third-degree AV block that is not expected to resolve or that persists at least 7 days after cardiac surgery
- Congenital third-degree AV block with a wide QRS escape rhythm, complex ventricular ectopy, or ventricular dysfunction
- Congenital third-degree AV block in the infant with a ventricular rate less than 55 beats/min or with congenital heart disease and a ventricular rate less than 70 beats/min
- Patients with congenital heart disease and sinus bradycardia for the prevention of recurrent episodes of intra-atrial reentrant tachycardia; SND may be intrinsic or secondary to antiarrhythmic treatment
- Congenital third-degree AV block beyond the first year of life with an average heart rate less than 50 beats/min, abrupt pauses in ventricular rate that are two or three times the basic cycle length, or associated with symptoms due to chronotropic incompetence
- Sinus bradycardia with complex congenital heart disease with a resting heart rate less than 40 beats/min or pauses in ventricular rate longer than 3 s
- Patients with congenital heart disease and impaired hemodynamics due to sinus bradycardia or loss of AV synchrony
- Unexplained syncope in the patient with prior congenital heart surgery complicated by a transient complete heart block with residual fascicular block after a careful evaluation to exclude other causes of syncope

Recommendations for Pacing After Cardiac Transplantation
- Persistent inappropriate or symptomatic bradycardia not expected to resolve and for other class I indications for permanent pacing

Recommendations for Pacing to Prevent Tachycardia
- Sustained, pause-dependent VT, with or without QT prolongation
- High-risk patients with congenital long-QT syndrome

Recommendations for Pacing in Patients with Hypertrophic Cardiomyopathy
- SND or AV block in patients with hypertrophic cardiomyopathy as described previously (see "Recommendations for Permanent Pacing in Sinus Node Dysfunction" and "Recommendations for Acquired Atrioventricular Block in Adults")

6.1.1 Contraindications

Pacemakers are generally contraindicated for the following applications [32]:
- ICD (especially when use of a unipolar pacemaker lead is intended; it may cause unwanted delivery or inhibition of ICD therapy).
- Use of the MV sensor for patients with an ICD.
- Use of the MV sensor in patients with only unipolar leads because a bipolar lead is required in either the atrium or the ventricle for detection of MV.
- Single-chamber atrial pacing in patients with impaired AV nodal conduction.
- Atrial tracking modes for patients with chronic refractory atrial tachycardia (atrial fibrillation or flutter), which might trigger ventricular pacing.
- Dual-chamber and single-chamber atrial pacing in patients with chronic refractory atrial tachycardia.
- Asynchronous pacing in the presence (or likelihood) of competition between paced and intrinsic rhythms.

6.2 Implantable Cardioverter-Defibrillator Implantation Indications

ICDs are intended to provide ventricular antitachycardia pacing and ventricular defibrillation shocks for automated treatment of life-threatening ventricular arrhythmias. Indications for ICD implants are based on results of large randomized clinical trials, which approved the effectiveness of this therapy for the secondary as well as the primary prevention of SCD [36, 37]. An overview of the most important clinical trials is provided below. However, ICD implantation is not indicated for patients whose VT may have a transient cause (acute myocardial infarction, electrocution, drowning) or a reversible cause (digitalis intoxication, electrolyte imbalance, hypoxia, sepsis) or patients who have a unipolar pacemaker [70].

According to ACC/AHA/HRS 2008 Guidelines for Device-Based Therapy of Cardiac Rhythm Abnormalities [34, 35], the indications (class I and class IIa) for ICDs are described in the following paragraphs.

- Survivors of cardiac arrest due to ventricular fibrillation or hemodynamically unstable sustained VT after evaluation to define the cause of the event and to exclude any completely reversible causes
- Structural heart disease and spontaneous, sustained VT, whether hemodynamically stable or unstable
- Syncope of undetermined origin with clinically relevant, hemodynamically significant sustained VT or ventricular fibrillation induced during electrophysiologic testing
- LV ejection fraction (LVEF) less than 35 % due to prior myocardial infarction in patients who are at least 40 days after the event and are in New York Heart Association (NYHA) functional class II or III
- Patients with nonischemic dilated cardiomyopathy who have an LVEF less than or equal to 35 % and who are in NYHA functional class II or III
- LV dysfunction due to prior myocardial infarction in patients who are at least 40 days after the event, have an LVEF less than 30 %, and are in NYHA functional class I
- Nonsustained VT due to prior myocardial infarction, LVEF less than 40 %, and inducible ventricular fibrillation or sustained VT during electrophysiologic testing
- Unexplained syncope, significant LV dysfunction, and nonischemic dilated cardiomyopathy
- Sustained VT and normal or near-normal ventricular function
- Patients with hypertrophic cardiomyopathy who have one or more major risk factor for SCD
- Prevention of SCD in patients with arrhythmogenic right ventricular dysplasia/cardiomyopathy who have one or more risk factors for SCD
- Long-QT syndrome in patients who are experiencing syncope, VT, or both while receiving β-blockers
- Nonhospitalized patients awaiting cardiac transplantation
- Patients with Brugada syndrome who have had syncope
- Patients with Brugada syndrome who have documented VT that has not resulted in cardiac arrest
- Patients with catecholaminergic polymorphic VT who have syncope, documented sustained VT, or both while receiving β-blockers
- Patients with cardiac sarcoidosis, giant cell myocarditis, or Chagas disease

Recommendations for Implantable Cardioverter-Defibrillators in Pediatric Patients and Patients with Congenital Heart Disease

- Survivors of cardiac arrest after evaluation to define the cause of the event and to exclude any reversible causes
- Patients with symptomatic sustained VT in association with congenital heart disease who have undergone hemodynamic and electrophysiologic testing; catheter ablation or surgical repair may offer possible alternatives in carefully selected patients
- Patients with congenital heart disease with recurrent syncope of undetermined origin in the presence of either ventricular dysfunction or inducible ventricular arrhythmias during electrophysiologic testing

6.2.1 Significant Clinical Studies of ICDs

Clinical studies deal with assessing the long-term clinical effect of treatment. In the early days of use of device therapy with ICDs, it was necessary to compare the effect of this new treatment with the existing medical therapy. Here, description of significant clinical studies will be divided into those investigating secondary and primary prevention. Interpretation of study endpoints and results is not possible here without major simplification because additional meta-analyses were performed for many studies, and clinical efficacy for various specific indications was observed.

Clinical studies in medicine are divided according to the following properties [38]:

- **Controlled/uncontrolled study**
  A controlled study involves assessment of treatment efficacy in comparison with an untreated group or a group treated with another drug or method. This group is referred to as the control group, and it significantly increases the informative value of the study. Placebo, another type of treatment, or results of standard therapy can all be used as controls.
- **Randomized/nonrandomized study**
  In a controlled study, it is necessary to ensure that patients as clinically similar as possible be represented in the groups. This is achieved by randomization of patients into treatment and control groups. In multicenter studies, it is

Table 6.3 Characteristics of randomized studies assessing ICD efficacy as part of secondary prevention [39]

Study	Years	Patients (n)	Main inclusion criteria	Reduction in mortality in the ICD group (%)
AVID	1993–1997	1,013	Resuscitation for VF, VT with syncope, LVEF≤40 %	31
CIDS	1990–1998	659	Resuscitation for VF, VT with syncope, VT with a cycle length≤400 ms, LVEF≤35 %	30
CASH	1987–1998	288	Resuscitation for VF	39

important to have an approximately equal number of subjects in the treatment and control groups at each center.

- **Prospective/retrospective study**
  A retrospective study investigates the effect of a phenomenon that was used in treatment in the past. Thus, it is rather an analysis and interpretation of data obtained previously. In a prospective study, the protocol, methods, and study subject characteristics are designed first; only then are subject enrollment and the actual study initiated.
- **Single-blind/double-blind study**
  When performing controlled studies, the result can be biased by patient or investigator expectations. Thus, the use of placebo makes sense only when the patient is unaware whether he or she is taking an active substance or placebo; such studies are referred to as single-blinded. If the investigator is also unaware whether the patient is treated with placebo or the active substance, the study is referred to as double-blinded. Studies in which even the team processing the data is unaware of which group is receiving which treatment are referred to as triple-blinded.
- **Single-center/multicenter study**
  When a large number of patients needs to be recruited, studies are conducted in multiple centers or worldwide. Results are not biased by regional differences in the treatment of enrolled subjects.

6.2.1.1 Secondary Prevention Studies

Clinical studies analyzing secondary prevention of SCD (Table 6.3) show a consistently higher efficacy of ICD treatment compared with antiarrhythmic drugs [39].

Antiarrhythmics Versus Implantable Defibrillator (AVID)

The AVID study was conducted in 1,013 patients with near-fatal ventricular fibrillation or sustained hemodynamically unstable VT and an LVEF<35 %. They were randomized into groups with pharmacological treatment with amiodarone or ICD implantation. In the pharmacological treatment group, 97.4 % of patients received amiodarone and 2.6 % received sotalol. The study was stopped early when a significant reduction in mortality – 38 % at 1-year follow-up and 31 % at 3 years – was shown in patients with an ICD compared with the group of patients taking an antiarrhythmic drug. A subgroup analysis revealed that patients with an LVEF<34 % benefited most from ICD. No benefit from ICD compared with amiodarone was found in patients with an LVEF>34 %. By contrast, patients with an LVEF of 20–34 % had a significantly reduced mortality with the use of ICD in years 1 and 2. A similar trend was shown in the group with an LVEF<20 %, but the results were not statistically significant because of the small number of patients.

Canadian Implantable Defibrillator Study (CIDS)

The CIDS study was conducted in 659 patient-survivors of ventricular fibrillation who had documented VT with syncope, sustained VT with presyncope or stenocardia, and an LVEF<35 %. They were randomized into groups with pharmacological treatment or ICD implantation. The study findings confirmed a 30 % reduction in mortality in the ICD group. A subgroup analysis identified patients who benefited most from ICD treatment (a 50 % relative risk reduction in all-cause mortality): age>70, LVEF<35 %, NYHA classes III and IV. These patients exhibited a 1-year mortality rate of 30 % when treated with amiodarone, but only 14 % with ICD treatment.

Cardiac Arrest Study Hamburg (CASH)

The CASH study was performed in 288 patients after cardiac arrest secondary to VT who were randomized into groups with pharmacological treatment or ICD implantation. At the 2-year follow-up, mortality from sudden death was 2 % in the group treated with ICD implantation (99 patients) compared with 11 % in the group of 189 patients treated with amiodarone or metoprolol. All-cause mortality was 12.1 % in the ICD group versus 19.7 % in the group treated with amiodarone or metoprolol, which means a relative reduction in all-cause mortality of 39 %.

6.2.1.2 Primary Prevention Studies

Multicenter Automatic Defibrillator Implantation Trial (MADIT)

The MADIT was conducted from 1990 to 1996 and included 196 patients with ischemic cardiomyopathy, an LVEF<35 %, unsustained VT, and sustained VT resistant to procainamide infusion during electrophysiologic testing. They were randomized into groups with pharmacological treatment with antiarrhythmic drugs or ICD implantation. The study was stopped prematurely. In the ICD group, only 16 % of patients died versus 39 % in the pharmacological treatment group (a 54 % relative risk reduction in all-cause mortality); of these deaths, cardiac causes accounted for 12 % and 26 %, respectively. A subgroup analysis revealed that patients with an LVEF < 26 % benefited most from ICD.

Multicenter Unsustained Tachycardia Trial (MUSTT)

The MUSTT study was conducted from 1990 to 1998 and included 704 patients with ischemic cardiomyopathy, an LVEF < 40 %, and asymptomatic unsustained VT. A total of 351 patients underwent electrophysiologic testing and the remaining 353 were only followed up. Patients with positive electrophysiologic testing were randomized to receive either no antiarrhythmic treatment or treatment with antiarrhythmic drugs indicated by electrophysiologic testing. Patients in whom antiarrhythmic drugs failed had an ICD implanted. Patients were followed up for a period of 5 years. At 5 years, all-cause mortality was 24 % in the ICD group, 55 % in the group treated with antiarrhythmic drugs, and 48 % in patients followed in the registry without treatment. A subgroup analysis showed that only patients treated with ICD benefited from electrophysiologic testing. Over a period of 5 years, 9 % of ICD patients suffered cardiac arrest or arrhythmic death compared with 37 % of patients treated with antiarrhythmic drugs and 32 % of patients in the untreated group. The MUSTT study supported the findings of the MADIT study.

MADIT II

The MADIT II was conducted from 1997 to 2001 and assessed the use of ICD without prior electrophysiologic testing in patients with an LVEF < 30 % and with an occurrence of fewer than ten ventricular ectopic beats or couplets per hour during Holter monitoring. Patients with VT were excluded. Patients were randomized into groups with pharmacological treatment or ICD implantation. The study was stopped prematurely. There was a 30 % reduction in mortality in the ICD group.

Sudden Cardiac Death Heart Failure Trial (SCD-HeFT)

The SCD-HeFT, conducted from 1997 to 2003, included a total of 2,521 patients who were randomized to receive placebo (n = 847), amiodarone (n = 845), or ICD implantation (n = 829). All patients had chronic heart failure corresponding to NYHA class II or III of ischemic or nonischemic etiology and an LVEF < 35 %. This study revealed that, at 5 years of follow-up, patients with an ICD showed an absolute reduction in mortality of 7.2 % compared with the control group of patients. At 3 years of follow-up, all-cause mortality was 17.1 %, 24.0 %, and 22.3 % in the ICD group, the amiodarone group, and the placebo group, respectively. At 3 years of follow-up, all-cause mortality was 28.9 %, 34.1 %, and 35.8 % in the ICD group, the amiodarone group, and the placebo group, respectively. ICD implantation reduced all-cause mortality by 23 % compared with placebo; the effect of amiodarone was not significant [42].

Defibrillators in Nonischemic Cardiomyopathy Treatment Evaluation (DEFINITE)

The DEFINITE study was conducted from 1998 to 2002 and included 458 patients with dilated nonischemic cardiomyopathy. They were randomized into groups with pharmacological treatment with antiarrhythmic drugs or ICD implantation. The results showed a significant reduction in mortality in the ICD group and a significant reduction in the incidence of sudden cardiac death [40, 43].

Comparison of Medical Therapy, Pacing and Defibrillation in Heart Failure (COMPANION)

The COMPANION study, conducted from 1997 to 2002, included 1,520 patients with NYHA class III and IV heart failure symptoms and with an LVEF < 35 %, a QRS width > 120 ms, and an LV end-diastolic dimension (LVEDD) > 60 mm, regardless of the etiology of cardiac dysfunction. They were randomized into groups treated with optimal pharmacological therapy (n = 308), optimal pharmacological therapy plus cardiac resynchronization therapy (CRT) with a pacemaker (CRT-P) (n = 617), and optimal pharmacological therapy plus CRT with a defibrillator (CRT-D) (n = 595). The study was stopped early because of a reduction in mortality by more than 20 % in the CRT-P arm in combination with the CRT-D arm compared with the arm with pharmacological therapy alone. A significant reduction in mortality in favor of CRT was even more marked in the CRT-D group (43 %) [41].

Defibrillator in Acute Myocardial Infarction Trial (DINAMIT)

The DINAMIT was conducted from 1998 to 2002 and included patients early after an acute coronary event, with an LVEF ≤ 30 % evaluated 3 days or more after developing myocardial infarction and a finding of reduced heart rate variability evaluated by the standard deviation from an average length of cardiac cycles in sinus rhythm. This open, multicenter, randomized study included a total of 676 patients who were randomized into groups with pharmacological treatment with antiarrhythmic drugs or ICD implantation. The conclusion drawn from the results of the DINAMIT was that ICD implantation failed to have an impact on all-cause mortality among the patients. Prophylactic ICD implantation significantly reduced (by 58 %) the mortality from sudden arrhythmic death; however, patients with ICDs had a significantly higher (by 78 %) mortality rate from other than arrhythmic causes during the follow-up period [42].

Coronary Artery Bypass Graft Patch Trial (CABG-Patch)

The CABG-Patch was conducted from 1992 to 1998 and assessed the effect of prophylactic ICD implantation in 900 patients with ischemic heart disease, LV dysfunction (with an LVEF < 35 %), and a finding of ventricular late potentials. Patients were randomized to receive either ICD or no treatment.

All patients underwent surgical revascularization; in addition, one group was randomly assigned to ICD implantation. During the follow-up period, no difference in all-cause mortality was found between the ICD group and that without ICD. During the follow-up period, cumulative mortality from arrhythmic death was 6.9 % and all-cause mortality was 21.1 % in the control group; in the ICD group, cumulative mortality from sudden death was 4.0 % and all-cause mortality was 22.8 %. Treatment with ICD implantation had no impact on all-cause mortality.

6.3 Indications for Cardiac Resynchronization Therapy

CRT-D and CRT-P function are indicated in patients with moderate to severe heart failure (NYHA class III/IV) in whom symptoms persist despite stable optimal pharmacological therapy and who remain with an LVEF ≤ 35 % and a QRS width ≥ 120 ms [44]. The stages of heart failure are described in Table 6.4.

There is no universally accepted definition of heart failure. The most common definition is a hemodynamic one. The term *chronic heart failure* thus refers to a number of symptoms that are caused by impairment of the heart's work when, in spite of sufficient ventricular filling, cardiac output drops and the heart is unable to meet the metabolic needs of tissue. To establish the diagnosis of chronic heart failure, symptoms must be present and impaired cardiac function must be demonstrated objectively. Heart failure is a syndrome, not a definite diagnosis.

The term *compensated heart failure* refers to a condition in which there has been a resolution of clinical signs and symptoms of heart failure because of compensatory mechanisms or treatment. The term *asymptomatic dysfunction* refers to a condition in which there is reduced systolic and/or diastolic LV function but the patient is without either treatment or symptoms. Chronic heart failure develops as a result of dysfunction of the ventricular myocardium, arising in a number of cardiovascular diseases. This dysfunction can be systolic, diastolic, or both. The leading cause of chronic systolic heart failure is ischemic heart disease, usually a condition that occurs after suffering a myocardial infarction. The second leading cause is dilated cardiomyopathy. The other causes are less frequent.

The guidelines suggest sufficient demonstration of the clinical efficacy of CRT from large, randomized studies of patients with a widened QRS complex (≥ 120 ms). In most patients, this disorder is accompanied by a dyssynchrony of mechanical contraction that can be improved by using CRT. Currently, there is no evidence supporting this indication in patients with heart failure and a QRS width < 120 ms. Similarly, no evidence exists that it would be possible to improve the long-term efficacy of CRT by selecting patients based on echocardiographic assessment or another diagnostic method. Indications for CRT-P and CRT-D implantations overlap to a certain degree. The most recent guidelines on SCD emphasize that, when indicating CRT-D as part of primary prevention, it is necessary to take into account the patient's expected survival. These guidelines state explicitly that the use of ICD for the purpose of primary prevention is indicated in patients with heart failure with severe LV dysfunction regardless of the underlying disease and in whom survival rates longer than 1 year can be expected.

According to actual ACC/AHA/HRS 2008 Guidelines for Device-Based Therapy of Cardiac Rhythm Abnormalities [34, 35], the indications (class I and class IIa) for CRT are described in the following paragraphs.

- In instances of an LVEF ≤ 35 %, a QRS duration ≥ 0.12 s, and sinus rhythm, CRT with or without an ICD is indicated for the treatment of NYHA functional class III or ambulatory class IV heart failure symptoms with optimal recommended medical therapy.
- In instances of an LVEF ≤ 35 %, a QRS duration ≥ 0.12 s, and atrial fibrillation, CRT with or without an ICD is reasonable for the treatment of NYHA functional class III or ambulatory class IV heart failure symptoms with optimal recommended medical therapy.

Table 6.4 New York Heart Association classification of stages of heart failure [44]

NYHA class	Definition	Examples
I (Mild)	No limitation of physical activity. Ordinary physical activity does not cause undue fatigue, dyspnea, or palpitations	Carry 11 kg up eight steps; carry objects weighing 36 kg; shovel snow spade soil; ski; play squash, handball, or basketball; jog or walk 8 km/h
II (Mild)	Slight limitation of physical activity. Comfortable at rest, but ordinary physical activity causes fatigue, dyspnea, palpitations, or angina	Sexual intercourse without stopping; garden; roller skate; walk 7 km/h on level ground; climb one flight stairs at a normal pace without symptoms
III (Moderate)	Moderate limitations of physical activity. Comfortable at rest; less than ordinary physical activity causes fatigue, dyspnea, palpitations, or angina	Shower or dress without stopping; strip and make a bed; clean windows; play golf; walk 4 km/h
IV (Severe)	Severe limitation of physical activity. Symptoms occur at rest; any physical activity increases discomfort	Cannot do any of the above activities

- CRT is reasonable in patients with an LVEF ≤ 35 % with NYHA functional class III or ambulatory class IV symptoms who are receiving optimal recommended medical therapy and who have frequent dependence on ventricular pacing.

Resynchronization therapy is not indicated in patients in NYHA functional class IV for whom there is no obvious chance to improve the prognosis or quality of life.

6.3.1 Significant Clinical Studies of CRT

A study of biventricular pacing was published in 1990 [45]. Initial clinical use was described in a pilot study in France in 1995. The results of these and other studies suggested improvement in some patients with chronic heart failure. Afterward, large clinical studies followed [40, 46, 47].

6.3.1.1 Vigor in Congestive Heart Failure (Vigor-CHF)

The Vigor-CHF study was a randomized, controlled study that was conducted from 1996 to 1998 and included 53 patients of NYHA functional class III/IV. The aim was to assess cardiac performance by using hemodynamic parameters obtained by invasive measurements and echocardiographic testing. An improvement in hemodynamic parameters was observed following the implantation of a biventricular system.

6.3.1.2 Pacing Therapies for Congestive Heart Failure Study (PATH-CHF)

The randomized, controlled, single-blind PATH-CHF study that was conducted from 1995 to 1998 included 42 patients with chronic heart failure secondary to dilated cardiomyopathy in NYHA functional class III/IV, with a QRS > 120 ms, who were still symptomatic despite adequate medical therapy. The aim was to assess the benefit of LV or biventricular pacing with an optimal AV delay. The patients were randomized to receive pacing therapy (LV or biventricular) for 1 month, followed by a 1-month period of no pacing therapy, and then crossed over to the other pacing mode for 1 month. Two dual-chamber pacemakers were used, each of which had its own sensing bipolar lead in the right atrium and a unipolar ventricular lead placed either endocardially in the right ventricle or epicardially in the left ventricle. The pacing modes for LV or right ventricular pacing and for biventricular pacing were VDD and VVT, respectively. The study results showed improvement with biventricular as well as LV pacing alone. Right ventricular pacing alone was shown to be inappropriate. The effect of AV delay and of lead placement was confirmed. Improvement occurred in all the parameters observed (NYHA class, questionnaire-assessed quality of life, oxygen consumption during exercise testing, and 6-min walk test).

6.3.1.3 InSync

The InSync study was an uncontrolled, multicenter clinical study conducted from 1997 to 1998. It initially included 103 patients. The aim was to assess the safety, success rate, and long-term effects of biventricular pacing. The inclusion criteria were chronic heart failure in NYHA class III/IV, a QRS width ≥ 150 ms, and significant dysfunction of a dilated left ventricle. Clinical improvement was observed in all the parameters assessed: the mean QRS complex width decreased from 180 to 156 ms; the average NYHA classification decreased from 3.3 to 2.2; the mean distance in the 6-min walk test changed from 295 to 347 m; and the quality-of-life questionnaire score changed from 55 to 35 points.

6.3.1.4 Ventak CHF/Contak CD

This study was started in 1998, published in 2001, and included 581 patients. They were randomized to either implantation of a Guidant Ventak CRT-D with an LV lead implanted via thoracotomy or implantation of Guidant Contak CRT-D with an LV lead implanted via the coronary sinus. The inclusion criteria were indications for ICD implantation, NYHA class II/III/IV, an LVEF < 35 %, and a QRS width > 120 ms. The patients were randomized to have biventricular pacing turned off or turned on for a period of 6 months; afterward, biventricular pacing was turned on in all patients. After 6 months, there was improvement in the questionnaire-assessed quality of life and the NYHA class. In the group receiving resynchronization therapy, there was an increase in the distance walked during the 6-min walk test and in maximal oxygen consumption.

6.3.1.5 Multisite Stimulation in Cardiomyopathy Study (MUSTIC)

The MUSTIC multicenter study was conducted from 1996 to 2000. The aim was to assess the clinical efficacy of biventricular pacing in patients with chronic cardiac insufficiency with a wide QRS complex. This study enrolled 131 patients, 76 of whom were followed up for 12 months. The group consisted of patients who had a sinus rhythm and no indication for pacemaker implantation because of arrhythmic causes (43 patients) or those with chronic atrial fibrillation who met the criteria for pacemaker implantation because of a slow ventricular response (33 patients). The patients were randomized to a 3-month period of biventricular pacing or no pacing, with a subsequent reciprocal exchange of both the modes. The study confirmed a significant increase in the distance walked during the 6-min walk test, an improved quality of life score, and increased maximal oxygen consumption.

6.3.1.6 The Multicenter InSync Randomized Clinical Evaluation (MIRACLE)

The randomized, controlled, double-blind MIRACLE study was conducted from 1998 to 2000. It included 266 patients. The inclusion criteria were chronic heart failure, an LVEF ≤ 35 %, an LVEDD ≥ 55 mm, NYHA class III/IV, and a QRS ≥ 130 ms. The aim of the study was to assess the change in the quality of life, NYHA class, and 6-min walk test following the implantation of a biventricular system. Patients who had received standard medication for more than 1 month underwent implantation and were randomized to the group with resynchronization therapy (n = 134) or the control group (n = 132) for a period of 6 months. Afterward, CRT was programmed to be turned on in all the patients. The 6-min walk test and questionnaire-assessed quality of life in the CRT group were evaluated as superior to those in the control group. In addition, there was an improvement in NYHA class and a reduction in LVEDD.

6.3.1.7 Multicenter InSync ICD Randomized Clinical Evaluation (MIRACLE ICD)

The MIRACLE ICD study, conducted from 1999 to 2001, included 639 patients who met the inclusion criteria from the MIRACLE study as well as indications for ICD implantation. After CRT-D implantation, the patients were randomized to the CRT group or the control group. The following parameters were assessed again: quality of life, changes in the NYHA class, 6-min walk test, echocardiographic parameters, and pharmacological medication. After 6 months, resynchronization therapy was activated in the control group. Improvements in the parameters observed were more significant in the group with active resynchronization therapy; however, no difference was found between the two groups in the 6-min walk test.

6.3.1.8 Comparison of Medical Therapy, Pacing and Defibrillation in Heart Failure (COMPANION)

See the description of this study in the "Secondary Prevention Studies" section included earlier in the chapter.

6.3.1.9 Cardiac Resynchronization in Heart Failure (CARE-HF)

This international, multicenter, controlled, single-blind study, conducted from 2001 to 2003, included a total of 813 patients from 83 centers in 12 countries. The aim of the study was to determine the effect of CRT on the morbidity and mortality of patients with moderate to severe heart failure. The inclusion criteria were heart failure persisting for more than 6 weeks, NYHA class III/IV despite optimal pharmacological therapy, an LVEF ≤ 35 %, LV dilatation as evidenced on echocardiography, and a QRS width > 120 ms. The patients were randomized into a group treated with pharmacotherapy and biventricular pacing (409 patients) and a group with pharmacological medication only (404 patients). In the CRT group, lower rates of hospitalization for cardiovascular causes and all-cause mortality were reported compared with the control group.

6.3.1.10 Resynchronization Reverses Remodeling in Systolic Left Ventricular Dysfunction (REVERSE)

This prospective, randomized, double-blind study, conducted in the period from 2004 to 2006, randomized 610 patients. Its aim was to assess whether CRT along with optimal pharmacological therapy can alleviate the development of heart failure compared with optimal pharmacological therapy alone. The inclusion criteria were mild heart failure of NYHA class I/II, a QRS width > 120 ms, an LVEF ≤ 40 %, and an LVEDD > 55 mm. The patients were randomized to optimal pharmacological therapy and CRT turned on (n = 419) or optimal pharmacological therapy and CRT turned off (n = 191). At 12 months, no statistically significant differences were found in the quality-of-life questionnaire or the 6-min walk test. An improvement in echocardiographic parameters was reported in the CRT group.

6.3.1.11 MADIT with Cardiac Resynchronization Therapy (MADIT-CRT)

Since 2005, this international, multicenter study included a total of 1,820 patients from 110 centers in 14 countries. It is the largest randomized CRT-D study of NYHA class I and II patients. The aim was to ascertain whether CRT-D implantation in high-risk patients with an LVEF < 30 %, a QRS > 130 ms, and mild heart failure of NYHA class I/II will reduce all-cause mortality or mortality from heart failure compared with ICD treatment alone. The patients were randomized into two groups in a 3:2 ratio (CRT-D vs. ICD). The study showed that early CRT-D implantation reduced the risk of mortality by 34 % compared with ICD implantation alone. Resynchronization also increased LVEF and reduced LVEDD.

The results of the above-mentioned studies, as well as others not described here, suggest a benefit from biventricular pacing. Still, according to some data, approximately 20–30 % of patients show no improvement following this therapy. These patients are referred to as nonresponders. Assessment of the efficacy of resynchronization therapy should be performed based on objective data obtained from echocardiographic or hemodynamic measurements and not only by assessing the NYHA functional class, which is subjective. However, which parameters are sufficient to assess the success of CRT is not yet fully clear.

7 Leads

Cardiac pacing or defibrillation leads are the most critical part of implantable systems. Because they ensure sensing of heart activity and transfer cardiac pacing pulses or defibrillation electric shocks from the implantable devices to the tissue, we can say that from the point of view of their efficiency, reliability, and long life, leads are more important than the devices themselves. Furthermore, they are implanted right in the heart's chambers either endocardially or epimyocardially; thus, compared with the devices implanted just under the skin or muscle, the technical or clinical failure of a lead represents a much higher risk for the patient.

The leads must comply with conditions for biological safety and materials used, and their construction must not cause any undesirable allergic or inflammatory responses or damage to tissue. Nowadays, the presumption is that the longevity of an implantable lead must be the patient's lifetime. Extraction of leads, especially in the case of defibrillation leads, is connected with high risk and might require cardiac surgical intervention (thoracotomy).

In the course of development, various kinds of leads were used. Historically, surface skin leads, esophageal leads for indirect cardiac pacing, or needle leads for direct cardiac pacing were developed. Later, epimyocardial leads were used that, on a limited scale and in justified cases, are still used today. For the time being, endocardial (intracardial) leads are used nearly exclusively. They are inserted via a venous route to the right ventricle or right atrium or via the coronary sinus to the coronary veins, where they are fixed using techniques described below. The advantage of endocardial leads is a quite simple method of their implantation that does not require any cardiac surgical intervention.

The lead and its accessories usually are sterilized using gas ethylene oxide before their final packing. Thus, the leads supplied by the manufacturer are already sterile and prepared for usage. Required temperature for their storage ranges from 0 °C to 50 °C [48].

7.1 Construction of Leads

Regarding the design of implantable leads, they consist of a fixing mechanism, pacing and shock electrodes, a conductor, insulation, and a connector (terminal). According to the clinical designation, we distinguish between atrial, ventricular, or left ventricular pacing leads, as well as defibrillation leads. The lead length ranges from 40 cm for the shortest atrial leads up to 100 cm for the longest left ventricular leads. The following lengths are considered standard: from 45 to 55 cm for the atrium, from 50 to 60 cm for the right ventricle, and from 75 to 100 cm for the left ventricle. A suitable length for a defibrillation lead ranges from 55 to 65 cm [49, 50]. The medical staff has to consider the optimal length of a lead during its selection. Certainly, it is undesirable to use a lead that is too short or too tight. However, length that exceeds optimal size by, for example, 40 cm, must be a superfluous strain on the patient.

Permanent assurance of excellent electrical performance of pacing is one of the most important requirements for cardiac pacing leads. However, the behavior of the cardiac pacing threshold in time shows two increases. First, it rises immediately after implantation of the lead in connection with local damage to the tissue; second, it rises gradually as a consequence of the inflammatory response of cellules and fibrosis created around the pacing electrode. Because of stabilization of the pacing threshold over time, modern leads usually are equipped with a steroid-eluting ring at the distal end. It is a ring made of porous silicone rubber that is filled by steroid. A nominal dose contains about 1.0 mg of dexamethasone acetate. The steroid is eluted from the ring slowly and gradually during its contact with body fluids. That way it depresses the inflammatory response that is considered to be a reason for the increase in pacing threshold that usually occurs in connection with implantation of the pacing leads.

D. Korpas, *Implantable Cardiac Devices Technology*, DOI 10.1007/978-1-4614-6907-0_7,

7.1.1 Fixation Mechanisms

The long-term operation of cardiac pacing is dependent on the stabile positioning of the pacing electrodes. Historically, the first way to fix the transvenous electrodes on the endocardial surface was called passive fixation. It dealt with various methods of attaching the fixation mechanisms, in the shape of small fillers or fixation tines (see Fig. 7.1), into the trabeculae of the right atrium or ventricle [4]. During implantation of the lead with passive fixation, its tines are captured immediately by the trabeculae; this can be confirmed by carefully pulling the lead back. Passive fixation is not technically demanding with regard to the implantation, especially into the apex. However, the tines of the passive fixation mechanism are covered quickly by fibrous tissue that makes later electrode repositioning or extraction more difficult or even impossible approximately 6 months after its implantation. The tines also increase the external diameter of the lead body. The porous electrode with passive fixation tines, which serves as a cathode for sensing and pacing, might have a considerably bigger area than the electrodes that are intended for different methods of fixation. Therefore the pacing threshold is lower.

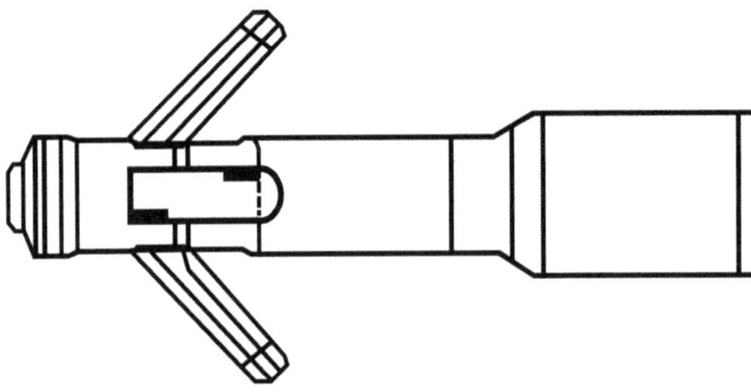

Fig. 7.1 Passive lead fixation

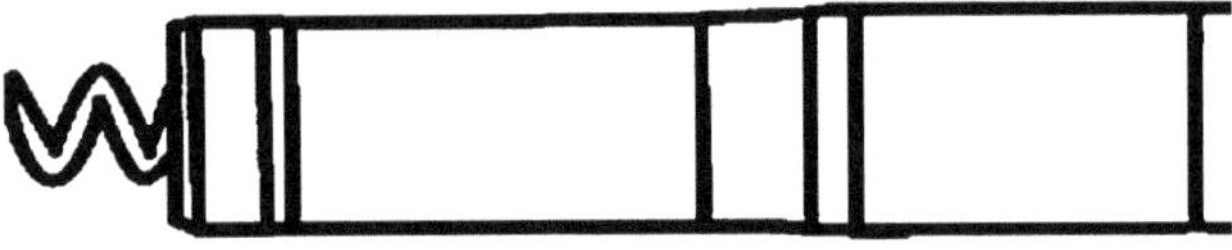

Fig. 7.2 Active lead fixation

Active fixation uses a principle of a helix that is extended into the endocardium (see Fig. 7.2). The helix can be retracted and the active fixation is released by that. The helix is extended/retracted by means of a mechanism that is joined to the lead conductor with a connector. The helix is extended by clockwise rotation of the connector pin and is retracted by counterclockwise rotation. Rotation is performed using a special fixation tool in the shape of a pair of small pliers. In the past, the helix was extended permanently and covered by, for example, polyethylene glycol so that it did not damage venous structures. This coating was dissolved in the blood bed after several minutes and the entire lead was screwed to affix the helix into the tissue. Construction of the extendable/retractable fixation helix anchors the distal electrode into the endocardium/myocardium without the support of the trabecular structures and offers various possibilities for lead positioning. If the extendable/retractable helix is electrically conductive and connected with the lead conductor, it serves as the cathode during pacing and sensing. There are also designs with a nonconductive helix and the electrode placed at the total distal end of the lead body. The leads are provided with fluorescence markers near the distal end. These markers do not transmit X-rays and they can be observed under skiascopy. They indicate when the helix is completely extended or completely retracted.

The left ventricular leads intended for pacing in the coronary bed are fixed in a different way. Because the pacing electrodes remain in the coronary veins, different invasive fixation cannot be considered. Therefore, the distal ends of the left ventricular leads are preformed either to a spiral curve (a pigtail) or to a suitable angled curve (*J*-curve); they are equipped with two very thin tines so that they generate a slight force on the narrow wall of the coronary vein to avoid return movement (see Fig. 7.3). During implantation of the lead, the preformed shape is restraightened by means of a guidewire or a stylet; not until after its extraction does the distal end of the lead recover its preformed shape.

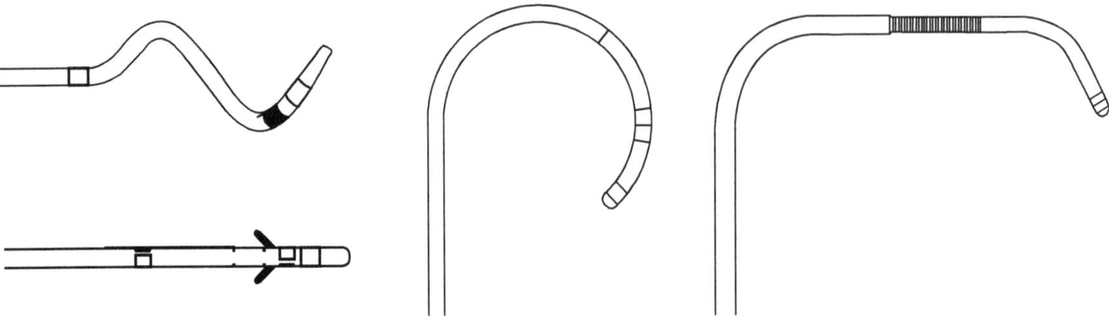

Fig. 7.3 Various left ventricular lead fixations

7.1.2 Construction of Pacing and Defibrillation Shock Electrodes

The pacing threshold is dependent on the current density created by the electrodes. The lesser the electrode diameter is, the higher the current density and the lower the pacing threshold are. In addition, the small surface of the electrode increases the contact resistance between the electrode and the tissue, which decreases the consumption of current from the source. On the other hand, per good sensing properties, a bigger electrode surface is better because it decreases sensing impedance and polarization. Although the issue of sensing impedance might be solved by appropriately setting the impedance of the input sensing amplifier, it is necessary to use electrodes with bigger surface area because of the polarization at the point of contact between the electrode and tissue. It was proved that it is possible to construct electrodes with a small diameter but a quite large surface. It can be achieved by creation of a porous texture (spatial microscopical ridging) at the pacing electrode, which is called fractal in some of the literature. In this way the electrode surface is enlarged while the diameter of the pacing electrode is kept the same. The textured surface minimizes polarization phenomena and increases pacing and sensing efficiency. Technologically it is achieved by sintering basic electrode material with platinum and iridium oxide or by constructing the electrode of microscopical metal fibers. A porous electrode made of platinum and iridium increases the active area and permanent stability of the electrode tine. For the time being, the most often used bipolar electrodes with active fixation have the area of the distal pacing electrode (the helix) – about 2 mm^2 – and of the proximal electrode (the ring), an area of about 40 mm^2. The relative distance of the pacing electrodes is dependent on the manufacturer, but it usually ranges from 10 to 20 mm.

Defibrillation shock electrodes are constructed as wound, small coils with minimal ascending, which ensures good mechanical elasticity. The winding is regarded as an integral area. Defibrillation leads can have the only defibrillation electrode (single-coil) or possibly two electrodes (dual-coil). The area of the distal shock electrode is about 310–450 mm^2; the area of the proximal shock electrode is about 480–660 mm^2 (the proximal electrode is always larger) [51, 52].

Because of the occasional need to extract cardiac pacing or defibrillation leads, it is necessary to minimize the possible ingrowth of fibrous tissue to the pacing and especially the defibrillation shock electrodes. Two approaches are used here. Either the area between individual fibers of the defibrillation electrode coil is injected with additional silicon insulation that restricts ingrowth of the fibrous tissue or the surface of the defibrillation electrode is coated with a polytetrafluoroethylene layer. It creates a porous structure with microscopic orifices that are too small for the fibrous tissue cells to penetrate, but they do not represent any difficulties for good electrical contact and, by that, good defibrillation. The leads must be constructed isodiametrically, which means with the same diameter along the entire length of the lead or with a decreasing diameter toward the distal end. It guarantees better patency through respective stenoses and makes eventual extraction simpler.

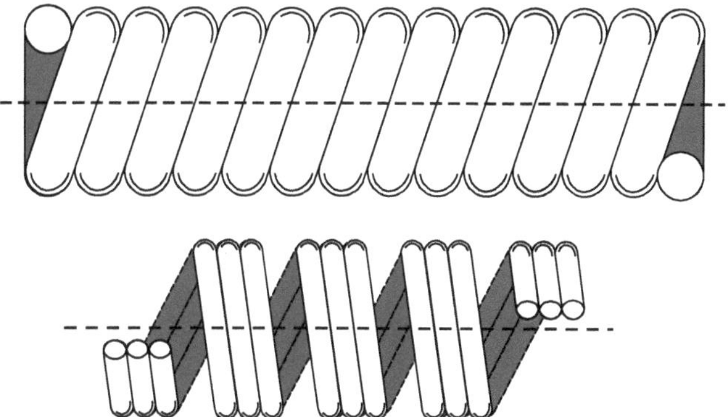

Fig. 7.4 Single-wound and triple-wound helix coil conductors

7.1.3 Construction of the Lead Conductor

To a large extent, the conductor designates mechanical properties of the whole lead and its good manipulability [4]. The conductor, together with its insulation, is the most mechanically stressed part of the entire implanted system. It is stressed by both the cardiac chamber afterloads and the patient's movements. At a pacing rate of 70 beats/min, it must withstand stress volume of 36 million cardiac afterloads per year. Mechanical properties of the leads depend on how it is wound internally. In terms of the arrangement of a bundle, one can distinguish between winding with a single-wound helix or multiwound helices (see Fig. 7.4). Winding with multiwound helices gives several advantages. First, a higher pitch of every helix, which the conductor consists of, ensures lower mechanical stress. Furthermore, the total electrical resistance is decreased by the parallel connection, and the eventual opening of one helix in the winding does not cut off the entire bundle.

Bipolar leads, which are used nearly exclusively nowadays, are distinguished by a coaxial or co-radial design of two bundles in the conductor (see Fig. 7.5). The bundles are insulated reciprocally by, for example, tetrafluoroethylene. There is an inner tubular channel winding through the inside of the conductor, which a stylet or a guidewire introduced while handling the lead.

The body of defibrillation leads generally contains the conductor for pacing and sensing with a co-radial structure of winding because the coaxial structure would further increase its diameter. It also contains one or two conductors

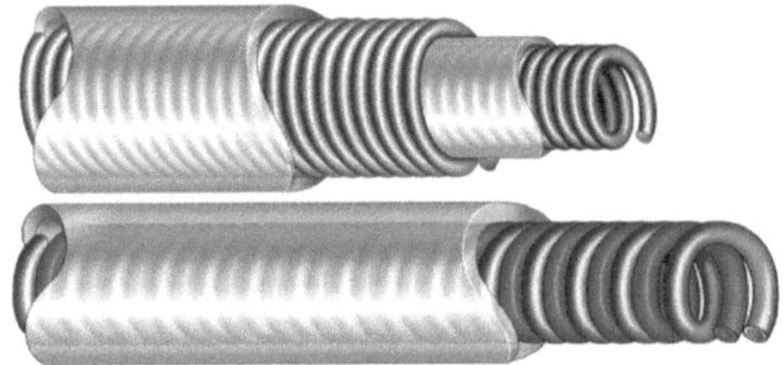

Fig. 7.5 Coaxial and co-radial lead conductor designs

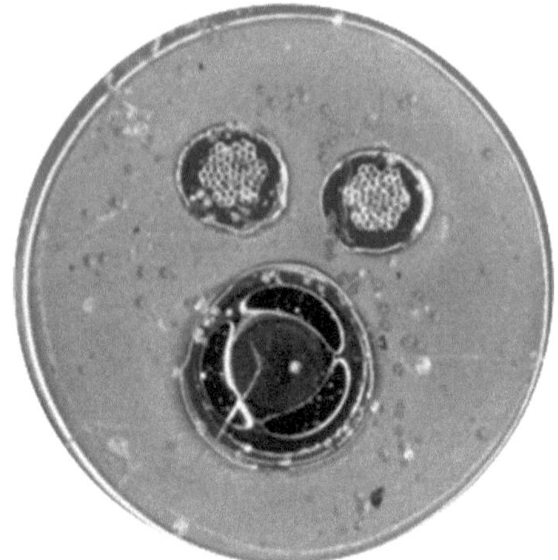

Fig. 7.6 Cross section of a defibrillation lead

for defibrillation, according to whether it deals with a lead with one or two defibrillation shock electrodes. Regarding their construction, defibrillation leads are divided into integrated bipolar leads, which include electrically connected electrodes (the proximal pacing electrode and the distal shock electrode), and separated bipolar leads, in which the electrodes are separated. The conductors are insulated by, for example, tetrafluoroethylene in isolated channels inside the body of the lead, which is made of silicone rubber. The second layer of insulation covers the lead body and ensures additional insulation and a unique diameter (see Fig. 7.6).

The most liable place for mechanical damage of the implanted lead is usually near the crossing of the subclavian vein and the first rib. Fixation that is too strong at the point where the lead enters into the vein might damage the insulation, especially if a suture sleeve (lead protector) is not used for the suturing.

The lead conductor, together with the pacing electrodes, predominantly designate the electrical properties of the lead. The total electrical resistance of the conductor between the connectors and pacing electrodes might range from about 30 to 150 Ω. It depends on the lead length. This value is not negligible, and it creates a considerable part of the impedance value within the pacing circuit. With pacing by a current of, for example, about 7 mA, there is a decrease in voltage up to 1 V just on the lead conductor. The resistance of defibrillation leads is much less because of the usage of conductors with a larger diameter; the maximum is about 2 Ω.

7.1.4 Construction of Lead Connectors

In the early stages of the cardiac pacing, the incompatibility of lead connectors and connector headers on a device was a technical problem. In those days, the necessary technical standards were neither created nor applied yet. Development went through various socket joints and bayonet systems, to 5- to 6-mm unipolar or bifurcate bipolar connectors, and finally to 3.2-mm bipolar connectors. These connectors already contained two double-tandem contacts for every pacing electrode. Lack of unity also dominated in the area of seal plugs and placement of liners. Some manufacturers left them at the lead connector, others in the port of the device's connector header. In this situation, manufacturers made a provisional agreement dealing with a voluntary standard for lead connectors and seal plugs and used a 3.2 mm connector (VS-1). Lasting problems with the acceptability of this standard finally were solved by the acceptance of the international technical standard for a bipolar connector of the pacing lead, designated as IS-1 (Fig. 7.7). It currently deals with the international standard *ISO 5841–3:2000 Implants for Surgery – Cardiac Pacemakers – Part 3: Low-Profile Connectors (IS-1) for Implantable Pacemakers*. The connector is tubular to enable introduction of the stylet or guide wire. The standard currently is used for cardiac pacing leads. Despite this standard, when heart resynchronization therapy was put into wider practice, a specific connector – type LV-1 – was designed for the left ventricular lead in which the seal plugs are placed in a port on the device header instead of on the lead connector. This enabled the catheter used to access the coronary sinus to be pulled over the lead connector in the case of dislocation of the left ventricular lead that was previously applied. However, when the proficiency in insertion of the left ventricular leads increased, the usage of this connector was terminated.

The international standard for the connector of one electrode of the DF-1 defibrillation lead is *ISO 11318:2002 Cardiac Defibrillators – Connector Assembly DF-1 for Implantable Defibrillators – Dimensions and Test Requirements*. According to whether it uses one or two shock electrodes, the defibrillation lead might be equipped with two (IS-1 + DF-1) or three (IS-1 + 2 × DF-1) connectors. The DF-1 connector is unipolar only and it is not tubular.

In connection with the miniaturization of the dimensions of implantable cardioverter-defibrillators and cardiac resynchronization therapy for defibrillators and with the reduction of the mechanical connection of leads to the device connector header, an effort was made to deal with the elimination of

Fig. 7.7 Type IS-1 connector

quite a voluminous part of the defibrillation lead at a place of the individual connectors branching. For that reason a new connector was developed, designated DF-4, according to the international standard *ISO 27186:2010 Active Implantable Medical Devices – Four-Pole Connector System for Implantable Cardiac Rhythm Management Devices – Dimensional and Test Requirements*. Usage of this connector also reduces the volume of the device connector header because up to three connectors were integrated into only one (IS-1 + 2 DF-1).

7.2 Lead Materials

An implantable lead is a heterogeneous corpus that is characterized by every part has different requirements about which materials are selected for use. When considering the materials required, the basic parts of an implantable lead, regardless of its purpose or positioning and apart from the fixing mechanism, are the pacing, defibrillation shock electrodes, lead conductor, insulation, and connectors for connection to the implanted device. Although the materials in question already have been well investigated, to provide a good understanding of the biomedical and biophysical context, some details about the selection of various materials must be mentioned here.

7.2.1 Materials for Pacing and Defibrillation Shock Electrodes

The pacing electrodes are located at a position most distant from the device and thus at the farthest position in the heart chambers. Their function is to transfer electrical energy to the tissue. As with the other parts, the pacing electrodes must be innocuous, especially regarding biological and mechanical properties, and they must not damage the paced tissue in any way. Noble metals, such as platinum and gold, and corrosion-resistant alloys comply with this condition. Silver has toxic effects and cannot be used.

From the point of view of the cardiac pacing function, it is important to be concerned with the polarization voltage [4]. Between the metal pacing electrode and the tissue, a double layer of electric charge carriers is created in which electron conductivity of the metals is transferred to ion conductivity of the tissue electrolytes. The double layer behaves electrically as a capacitor, with quite a high capacity that is dependent on the material use for the pacing electrodes. This stray capacity is charged by the current of the cardiac pacing pulse, and it influences voltage behavior at the pacing electrode. The voltage created at the electrode, which is necessary for overcoming of the charged double layer, is called the polarization voltage. Corrosion-resistant steels and alloys dispose of the highest polarization voltage; gold has a lower polarization voltage and platinum even less. Silver compounds dispose of the lowest polarization voltage. However, they cannot be used because of their toxicity, as stated above.

Apart from suitable mechanical properties, another requirement for manufacturing is the resistance to corrosion in the aggressive environment of body fluids. Some alloy components dissolve or are corroded by the influence of polarization during the current flow. From this point of view, platinum is considered to be the best. When iridium is added, the allow also has suitable mechanical properties.

Nowadays, platinum and iridium alloys are used exclusively for the electrodes of implantable leads because the added iridium improves the chemical properties of the platinum. The alloy is even more chemically stable against the influence of various environments. The distal pacing electrode (tip) usually has higher platinum content in the alloy (90 % Pt, 10 % Ir), whereas the content of platinum in the proximal pacing electrode (ring) is lower (80 % Pt, 20 % Ir). Some types of electrodes are coated with iridium oxide. The size of the electrically active area of the pacing electrodes is dependent on the type of electrode (passive/active fixation, unipolar/bipolar, distal/proximal).

The distal shock electrode serves as the anode for sensing and pacing (for the integrated defibrillation leads) and as the cathode or anode for delivery of the electric shock during cardioversion or defibrillation. The proximal shock electrode serves as the anode or cathode for the delivery of electric shock during cardioversion or defibrillation. The shock electrodes also enable sensing of the electrograms that are used to monitor and evaluate the rhythm's morphology. The shock electrode is made of a wound wire that creates a coil around the lead body. Requirements for functionality and safety are similar to those of pacing leads. In addition, there is a necessity to withstand a high current density and to avoid the intergrowth of tissue around and among fibers of the shock electrode. The tissue ingrowth might make later extraction impossible and might influence the electric parameters of the system. Some manufacturers use the platinum and iridium alloy (80 % Pt, 20 % Ir) for the shock electrodes. However, others apply drawn brass wire instead. To avoid ingrowth of tissue, the coiled electrodes might be injected by a suitable material, for example, expanded polytetrafluoroethylene.

7.2.2 Lead Conductor Materials

The lead conductor transmits sensing and transfers the pacing pulses and defibrillation shocks from the device to the pacing or defibrillation electrodes. Triple or quadruple wound helices made of MP35N material are used most often. MP35N is the registered trademark for a special alloy of cobalt, chromium, and nickel. Other nickel alloys including a tantalum or silver core, called drawn-brazed-strand, also are used. The specific resistance of the materials tends to be in a degree of tens of an ohm per centimeter. The resistance of the entire lead conductor amounts to several tens of an ohms and rarely more than a hundred ohms. Bipolar leads use, for example, ethylene-tetrafluoroethylene [40] for insulation of individual bundles of the co-radial or coaxial winding.

7.2.3 Insulation Materials

The leads are located in the aggressive environment of body fluids, so the insulation materials are required to meet high demands with regard to their imperviousness, elongation at rupture, reaction to blood clotting, and long-term mechanical and chemical stability. Some materials might also require retentivity. In the past, different materials were tested, for example polyamide (nylon), polyurethane, polyethylene, or silicon rubber. Nowadays, silicon rubber and polyurethane predominantly are used for lead conductors' insulation. Silicon rubber is chemically stable in a body and it does not cause any inflammatory or rejection responses. However, this material is sticky when touched, and handling two leads made of this material in a vein is quite difficult. But this discomfort can be reduced by various kinds of surface coatings on the silicon insulation. Polyurethane has a good resistance against tearing and it is not so sticky. Polyurethane 55D as well as 75D or 90A are used most often, especially for the parts that are under higher mechanical stress. It deals with the connectors for interconnection with the device, suture sleeves, and mechanical fixation elements. The lead insulation made of 80A polyurethane was shown to be insufficient because of microscopic cracks in the surface. Polyurethane is also subject to degradation by metal ion oxidation, especially in the case of silver chlorides contained in drawn-brazed-strand conductors.

7.2.4 Lead Connector Materials

The point of the lead connection to the implanted device is under a high strain with regard to mechanical as well as electric stress. Connectors and the device are connected by means of setscrews with defined pressure. The materials used for the setscrews (device contacts) and lead connectors must be the same, otherwise electrolytic corrosion might occur during penetration of body fluids into the system. The pacing connectors use corrosion-resistant steel, for example, surgical steel designated 316L. For insulation of individual elements of the contact system, which should create waterproof interconnection, polyurethane is used.

7.3 Cardiac Pacing Leads

Cardiac pacing leads serve for pacing and sensing of the right atrium, ventricle, or both (see Fig. 7.8), and they must be compatible with the device used. The usage of endocardial leads is contraindicated for patients who are hypersensitive to the steroid contained in the lead tine, those with severe tricuspid valvular disease, and patients with mechanical tricuspid heart valves. Basic requirements for the endocardial leads are a small diameter, good mechanical flexibility, reliable fixation, and long-term electric stability.

The leads are supplied with stylets, fixation tools, a stylet introducer, a vein pick, a lead end-cap, and documents in the external and internal sterile bowl, both of which can be opened by tearing the packaging. Various types of stylets are supplied. They might be soft or firm, and straight or preformed to a *J*-shape for implantation to the atrium. A straight stylet might be introduced to the supplied lead in advance. The stylet introducer serves to simplify insertion of the stylet into the lead. The suture sleeve is an adjustable reinforcement in the shape of a small tube that is positioned around the external insulation of the lead. It serves for the secure fixation of the lead and its protection where suturing occurs. The usage of the suture sleeve decreases risk of damage caused by suturing directly across the lead body.

Fig. 7.8 Endocardial pacing lead

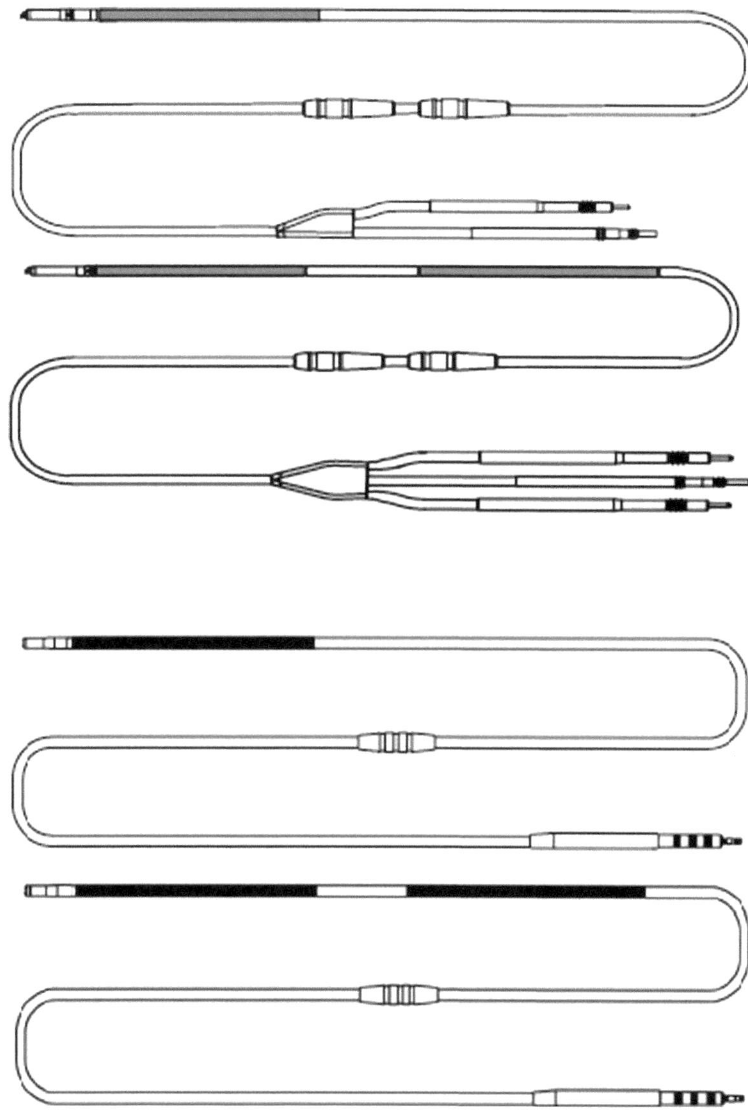

Fig. 7.9 Defibrillation leads – single coil DF-1, dual coil 2 × DF-1, single-coil DF-4, dual-coil DF-4 [51] (© 2012 Boston Scientific Corporation or its affiliates. All rights reserved. Used with permission of Boston Scientific Corporation)

7.4 Defibrillation Leads

Defibrillation leads ensure pacing and sensing and apply electric shocks during cardioversion and defibrillation of the automatic implantable defibrillation systems. Basic types of the leads are equipped with one or two shock electrodes (see Fig. 7.9). According to their construction, we distinguish the distal shock electrode from the electrically integrated or separated (dedicated, true bipolar) proximal pacing electrode. The usage of defibrillation leads has the same limitations as of the use of endocardial pacing leads.

7.5 Left Ventricular Leads for Cardiac Resynchronization Therapy

The usage of left ventricular endocardial leads instead of the originally used epimyocardial leads helps to decrease risk for the patient undergoing implantation. It also shortens the time of hospitalization and reduces total morbidity connected with the intervention. The basic requirements for the left ventricular leads are a flexible and mechanically safe distal tine, a suitable diameter, the ability of the lead body to bend

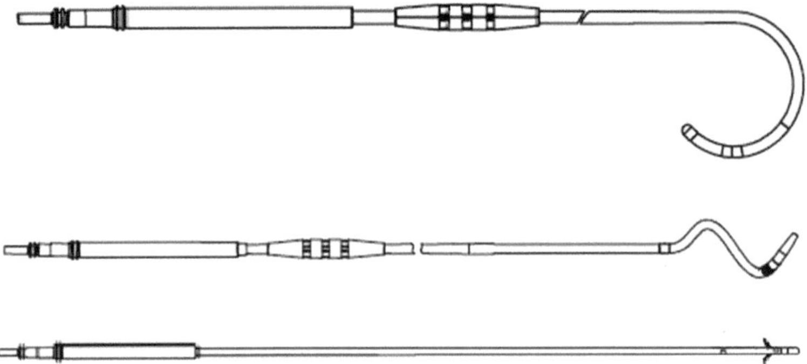

Fig. 7.10 Various left ventricular endocardial leads

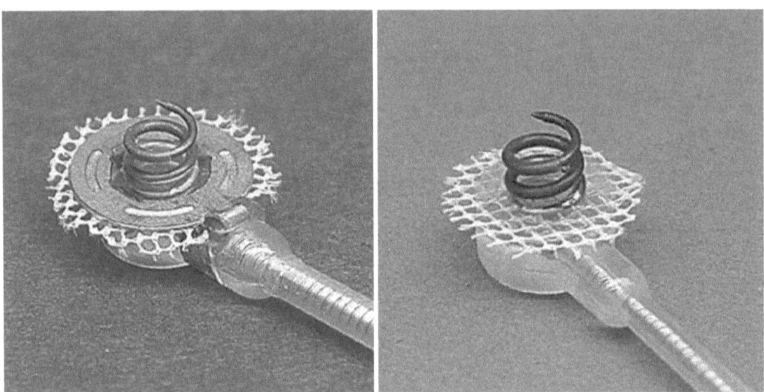

Fig. 7.11 Bipolar and unipolar epimyocardial leads

corresponding to anatomical circumstances, the possibility of rotation by the entire lead body, resistance to spontaneous torsion, and good visibility under a skiascope.

Nowadays, there are lots of leads available on the market for implantation via the coronary sinus (Fig. 7.10). Modern leads are positioned by means of the stylet or angiological wire (over-the-wire). In the optimal case, the lead offers both possibilities for implantation. Selection of a particular lead is made with regard to the anatomical arrangement of the coronary veins. Thicker leads with a preformed shape, enabling good fixation at the required position, are suitable for larger branches with little curved distance from the trunk [53, 54].

The diameter of the body of bipolar left ventricular leads is usually about 2 mm. However, at the distal end it decreases to 1.8 mm. Currently, the thinnest left ventricular leads have a lead body diameter of 1.5 mm, with a diameter of 1.35 mm at the distal end [55–57]. The way left ventricular leads are fixed does not require the distal pacing electrode to be completely at the end; within the left ventricular lead, both electrodes are shaped as "rings" and there is no contact with tissue at the end of the lead, which is placed in the vein. Depending on the design, the surface of pacing electrodes ranges from about 5 to 8 mm^2. The steroid-eluting ring and X-ray contrasting markers also are placed at the distal part of the lead.

7.6 Epimyocardial Leads

Permanent epimyocardial leads are used in cases when it is impossible to introduce the endocardial lead via the venous route, for example, in the case of central veins emphraxis, congenital heart disease, and repeated infections, or if positioning of the left ventricular lead for cardiac resynchronization therapy failed. Epimyocardial leads are screwed to the epimyocardium of the atrium or ventricles (see Fig. 7.11) and are sutured on its surface.

Proper fixation of the lead is important for keeping its excellent long-term electric parameters. Loosely fixed leads might move excessively or they might sense insufficiently, irritate epicardium, and cause higher threshold values. The intervention is usually performed during the cardiac surgery, most often using laparoscopy. It enables the use of various surgical approaches, such as the subxiphoid approach, left-sided thoracotomy, median sternotomy, transsphenoidal approach, or transmediastinal approach. The lead should be positioned and fixed at the avascular area without infarction, fat, and fibrosis. Before implantation, it is possible to use the epimyocardial lead for mapping by positioning the electrode in the epicardium [58].

These leads often lack steroid elution, and pacing thresholds are usually higher than that of the endocardial leads. It is tolerated that electric parameters are worse by 50 %; this means that the pacing threshold is 1.5 V and the amplitude of the sensed R wave is 3 mV. Epimyocardial leads are contraindicated for patients with a myocardium changed by fibrosis or for patients who went through a heavy myocardial infarction. In addition, if there is excessive epicardial fat, the results of the pacing cannot be satisfactory. If the lead will be connected to the device later or if it is put out of operation, the connector pin must be covered by the end-cap [59].

7.7 Subcutaneous Array Leads

Subcutaneous array leads are additional defibrillation leads implanted to the subcutis at the left lateral side of the chest. They improve distribution of the shock vector what covers the capacity of the heart ventricles better. The implantation of the additional lead represents a solution for patients with high defibrillation thresholds. The lead is created by one or several electrically active shock electrodes (see Fig. 7.12) that are connected to the implantable cardioverter-defibrillator by various ways, shown in Fig. 7.13. An example of a subcutaneous lead is the SQ array lead. The length of the individual shock electrodes is approximately 20 cm; the total length is 70 cm. The surface area of the shock electrodes reaches nearly 4,000 mm^2. We can expect that the decrease in the defibrillation threshold is about 30 % [60].

The subcutaneous array lead might cause an increase in the energy needed for cardioversion or defibrillation of the heart by means of the external defibrillator. Therefore, it is necessary to pay attention to the fact that the external defibrillator leads must not be positioned directly on the electrodes of the subcutaneous array leads. A decision about which connection should be used for the SQ array interconnection must be based on the demonstration of adequate safety reserves under the programmed shock energy by means of defibrillation threshold testing. Furthermore, the interconnections require usage of unipolar and bipolar adaptors.

7.8 Connectors and Adaptors

Over a period of several decades of development of permanent cardiac pacing and implantable defibrillation, many types of lead connectors were launched on the market. They are not always compatible with the ports used in device connector headers. Therefore, in practice, the usage of adaptors or universal adaptor sleeves might be required. In Tables 7.1, 7.2, and 7.3, various types of lead connectors and adaptors are shown [60, 61].

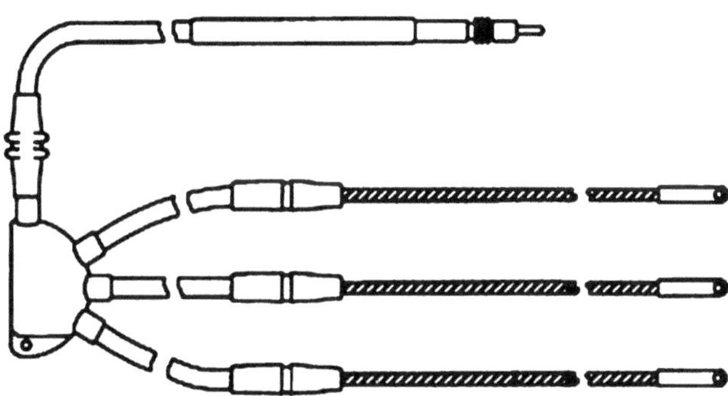

Fig. 7.12 SQ array [60] (© 2012 Boston Scientific Corporation or its affiliates. All rights reserved. Used with permission of Boston Scientific Corporation)

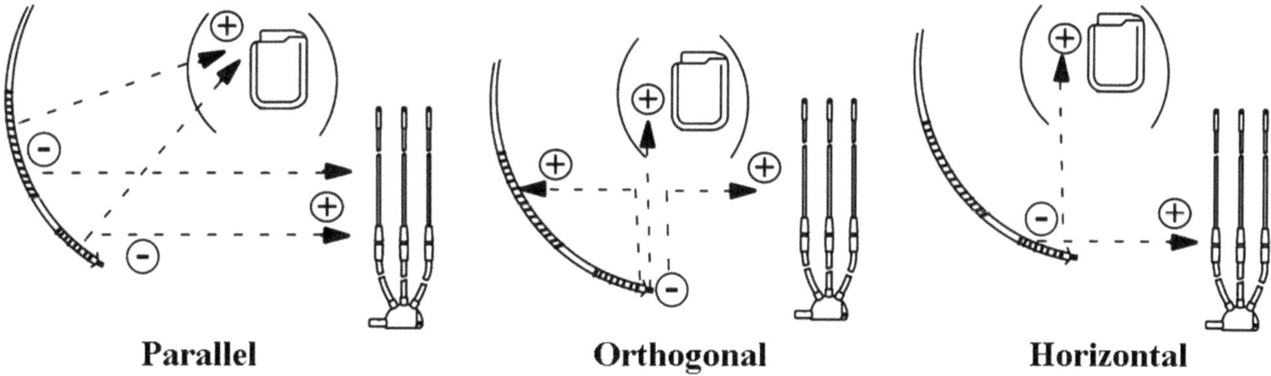

Fig. 7.13 SQ array configurations [60] (© 2012 Boston Scientific Corporation or its affiliates. All rights reserved. Used with permission of Boston Scientific Corporation)

Table 7.1 Lead connectors [60]

Connector illustration	Description
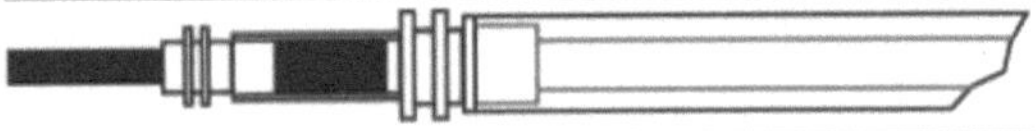	3.2-mm Cordis bipolar
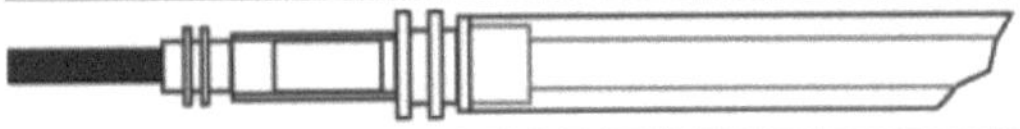	3.2-mm Cordis unipolar
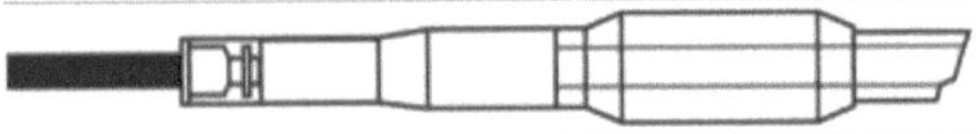	3.2-mm Teletronics unipolar
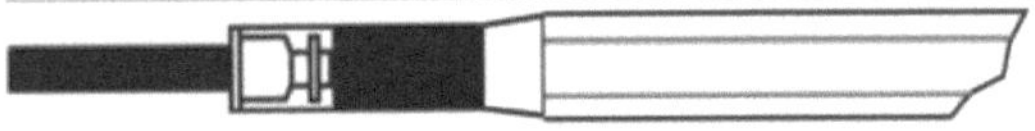	3.2-mm bipolar
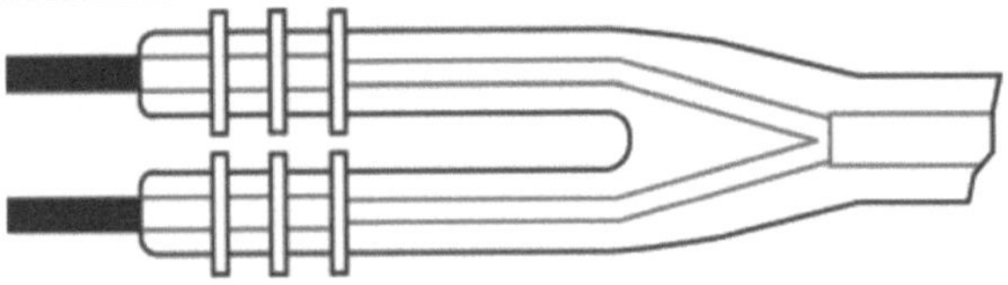	5-mm bifurcated bipolar
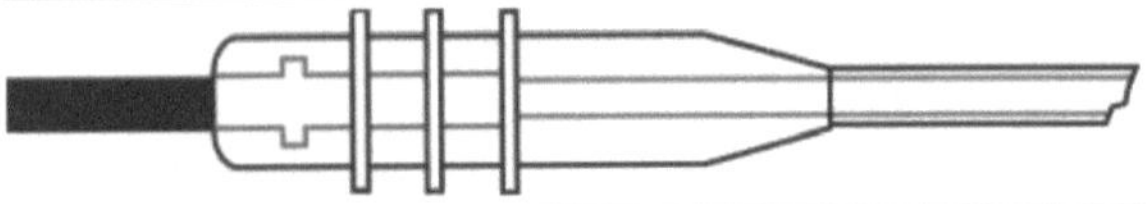	5-mm unipolar
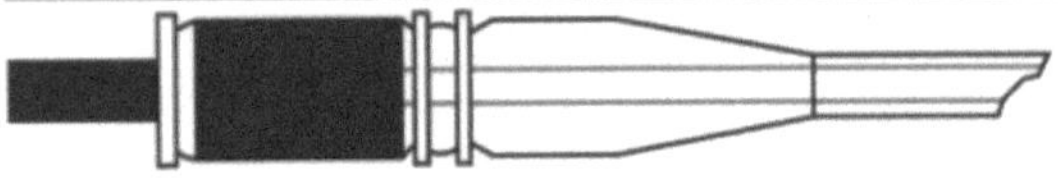	6-mm bipolar
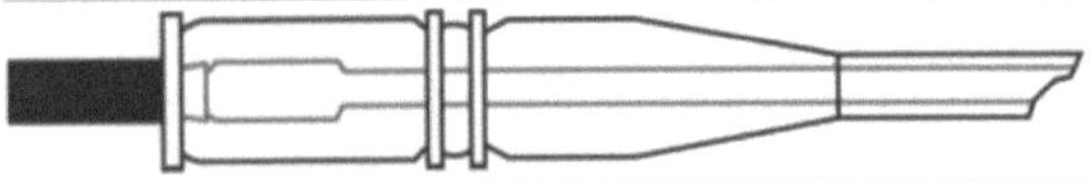	6-mm unipolar Cordis
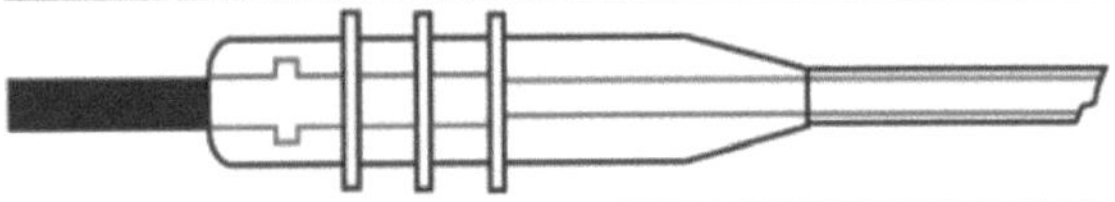	6-mm unipolar
	Tripolar
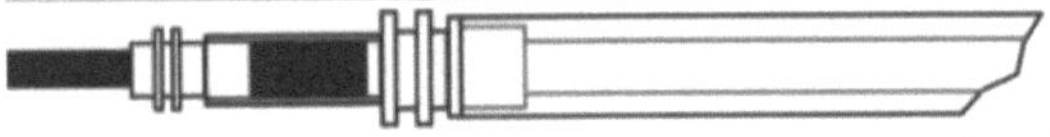	IS-1 bipolar
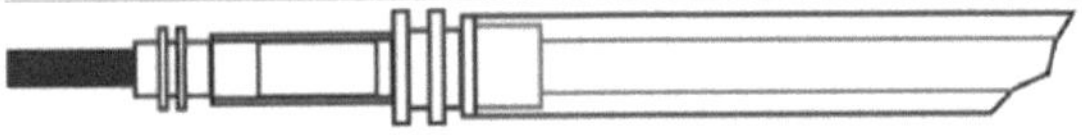	IS-1 unipolar
	LV-1 bipolar
	6.1 mm (defibrillation)
	DF-1 (defibrillation)
	DF-4 (defibrillation + pace)

Table 7.2 Examples of pacemaker headers [60]

Header illustration	Description
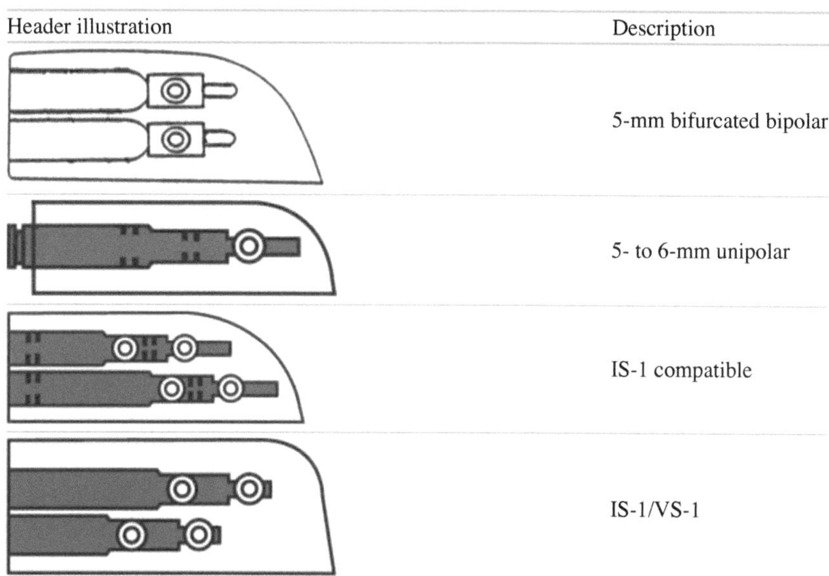	5-mm bifurcated bipolar
	5- to 6-mm unipolar
	IS-1 compatible
	IS-1/VS-1

Table 7.3 Lead connector plugs and adapters [60]

Plug/adapter illustration	Description
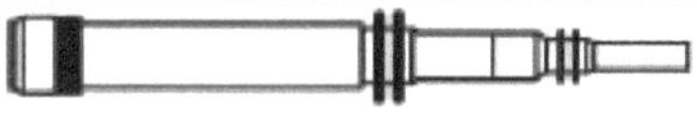	IS-1 plug
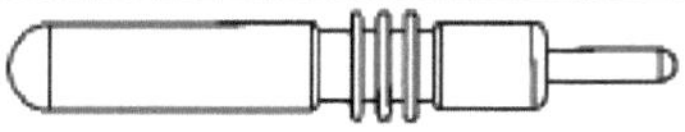	DF-1 plug
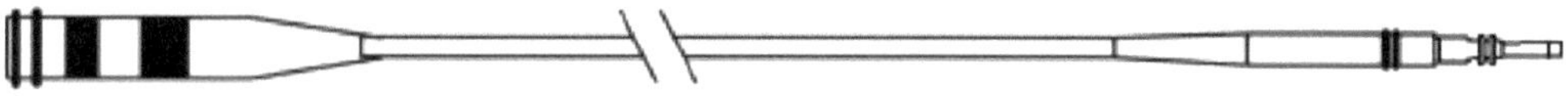	Extender IS-1
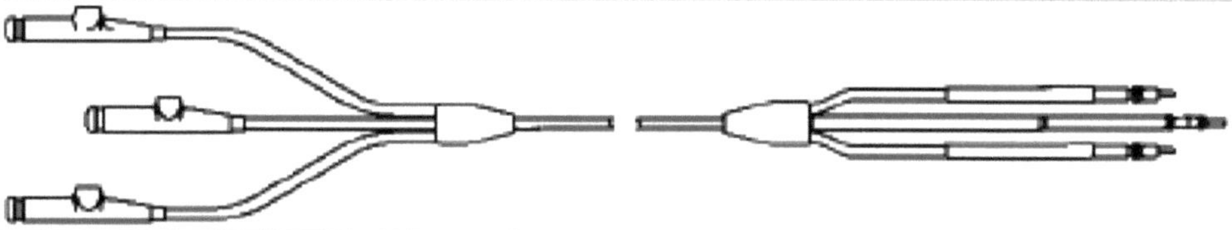	Extender 2×DF-1, 1×IS-1
	Adapter IS-1 – 2×4.75 mm
	Adapter IS-1 – LV-1
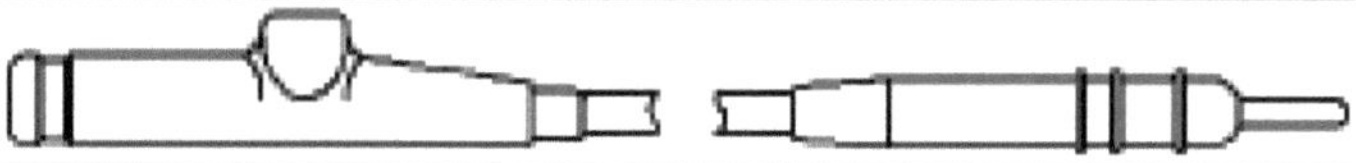	Adapter DF-1, – 6.1 mm
	Adapter DF-1, 6.1 mm
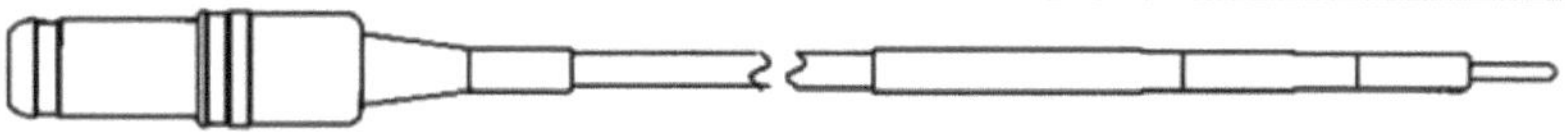	Adapter IS-1 unipolar, 6 mm
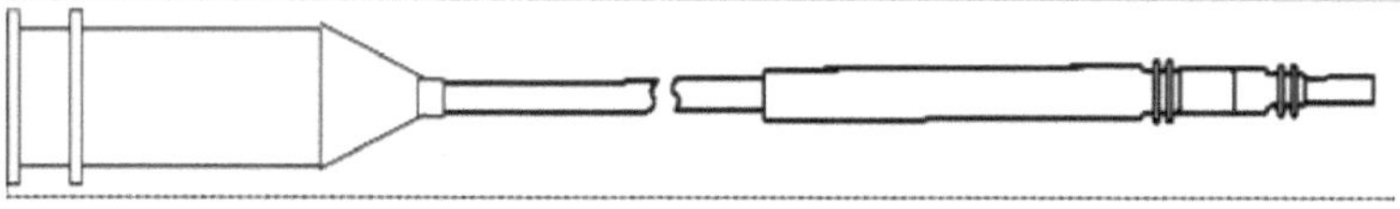	Adapter IS-1, 6 mm inline
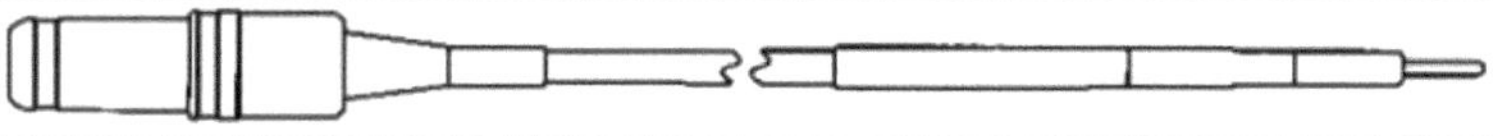	Adapter IS-1, 5-mm unipolar

8 Pacing Systems

A pacing system is defined as a medical device used for treatment of bradycardia or heart failure; it comprises the device itself and implanted leads. The pacemaker function is also incorporated into implantable defibrillators and biventricular pacemakers or defibrillators, which are, in a sense, considered to be superior devices. Most of the special functions of these devices will be dealt with in individual sections. Figure 8.1 shows a general block diagram of an implantable dual-chamber pacemaker. Up-to-date systems are controlled by a microprocessor, often with automatic gain control of input amplifiers.

8.1 Pacing Configuration

Fundamentally, two types of pacing or sensing configurations are available, according to the types of pacing leads: unipolar and bipolar (Fig. 8.2). This designation is rather misleading because both types contain an anode and a cathode, and both make use of one pacing electrode (cathode) in contact with tissue. The difference is in the placement of the second pacing electrode (anode) [4]. Pacing and sensing configurations are usually separately programmable for the atrium and the ventricle, and secure algorithms are included in the systems to prevent unintentional programming of a bipolar configuration when a unipolar lead is used. Up-to-date systems are capable of detecting the connection and diagnosing the lead type automatically.

If a unipolar pacing configuration is programmed, the pacing pulse will be applied between the cathode on the lead distal pacing electrode and the anode placed extracardially on the pacemaker can. The entire device or a certain limited area can be used as an indifferent pacing electrode. In a unipolar pacing configuration, pacing artifacts will be clearly visible on the surface electrocardiogram (ECG), which may help with interpretation. In a unipolar configuration, however, the anode is placed close to large muscles, which more often causes muscle stimulation or sensing of noncardiac biosignals (myopotentials). In this sensing configuration, the pacemaker may generally discern intrinsic cardiac signals with lower amplitudes, but it is also more susceptible to external electromagnetic interference.

In a bipolar configuration, the pacing pulse is applied between proximal and distal electrodes. Thus the anode is also placed in a heart chamber. In a bipolar sensing configuration, the sensitivity to signals occurring close to the electrodes is higher because of the relatively shorter distance between the lead electrodes. As a consequence, the pacemaker will less probably sense myopotentials and other signals unrelated to heart depolarization. A bipolar configuration provides higher pacing circuit impedance (because of the use of two lead conductors) and a slightly higher pacing threshold.

Because of the automatic setting of the sensitivity of the defibrillator sensing circuit, a large unipolar pacing pulse may avert sensing of weak intracardial fibrillation signals. For this reason, in implantable cardioverter-defibrillators or pacing systems with an additionally implanted defibrillator, bipolar configuration is used solely for the right atrium and ventricle. As far as left ventricular leads for cardiac resynchronization therapy are concerned, not only common bipolar or unipolar configurations are available. Based on the type of the lead and device, extended bipolar combinations of pacing electrodes can be opted for because the right ventricular proximal electrode can be utilized.

8.2 Pacing Impedance

Pacing impedance refers to the total value of pacing circuit impedance measured on output contacts of an implanted device. This value is important partly for long-term monitoring of the lead condition and partly for energy consumption. A rapid increase in the impedance indicates possible fracture of the lead conductor or another conduction pathway defect; a sudden drop, on the other hand, may signal damage to the

D. Korpas, *Implantable Cardiac Devices Technology*, DOI 10.1007/978-1-4614-6907-0_8,

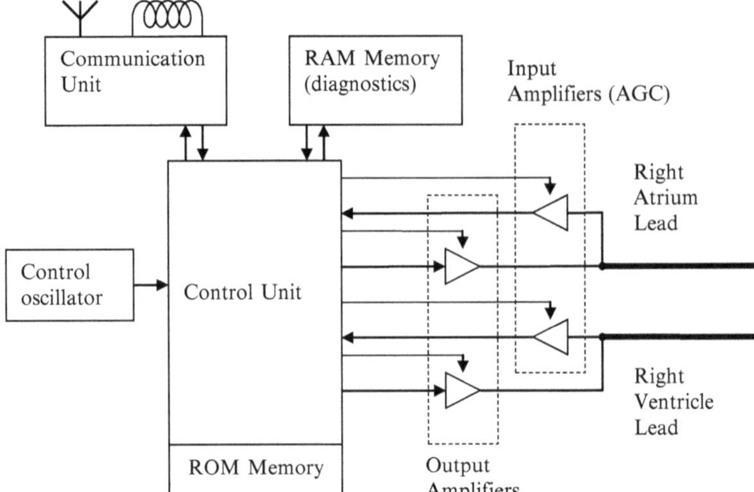

Fig. 8.1 Block diagram of a dual-chamber pacemaker

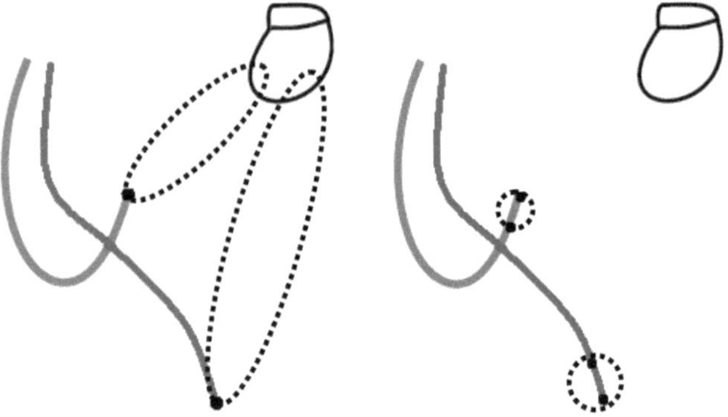

Fig. 8.2 Unipolar and bipolar configurations

lead insulation. Because the devices are powered by batteries with limited capacity, it is desirable to pace toward high impedance, which will limit the output current to a determined voltage.

The total impedance is influenced by several factors occurring on the pacing current pathway: the pure ohmic resistance of the lead conductor, lead electrode, and the myocardium themselves; and the polarization determined by gathering charges of opposite polarity on the pacing electrode and myocardium tissue interface. The lead conductor resistance causes a decrease in the voltage along the lead and partial heating of the lead. This poses an unnecessary loss of energy; therefore, the lead conductor resistance should be minimized. On the other hand, there should be strong resistance in the lead electrode to minimize flowing current and save the energy source. The pacing electrode resistance also depends on its geometric arrangement – a smaller surface provides higher resistance – but also higher current density and thus lower pacing threshold. The polarization impedance is the last component; it is determined by the movement of charged ions in the myocardium to the pacing electrode cathode. During the flow of pacing current, the cathode attracts positively charged ions and repels negative ions from the extracellular space. The cathode quickly surrounds itself mainly by sodium ions, while negative ions (e.g., chloride) are distant. Thus, two layers of oppositely charged ions are formed in the myocardium, and the current flow is caused by their movement. Capacity impedance occurs as a consequence of the two ion layers. This capacity effect grows during a pacing pulse with the peak on the trailing edge, and it avoids charge movement in the myocardium. Then, this increases the voltage required for pacing. The polarization impedance is directly proportional to the width of the pacing pulse and inversely proportional to the area of the lead electrode, which is why pacing with the shortest possible pulse and optimizing the surface of pacing electrodes, as described in Chap. 7, is recommended. With a lead implanted in situ in the endocardium, the system impedance is between approximately 200 and 2,000 Ω. Parasitic capacities, inducing a polarization effect in the direction opposite that of the pacing voltage, constitute an imaginary component of the impedance [40].

8.3 Basic Pacing Parameters

In the absence of intrinsic cardiac activity, the device paces the heart at a certain rate determined by a parameter referred to as the lower rate limit (LRL). The parameter is given in pulses per minute (see Chap. 9). LRL is also related to other pacing parameters, and even to the tachycardia zone setting. Certain systems allow decelerated pacing at night, that is, when it is assumed the patient is asleep and the heart rate decreases to the lowest level. At the beginning of the night mode, the pacing gradually decreases to the lowest level, and at the end, it gradually increases again to the daytime level.

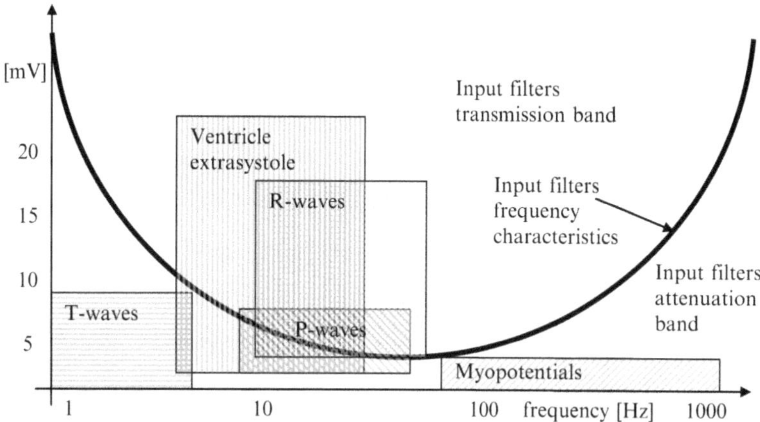

Fig. 8.3 Amplitudes and frequency ranges of sensed biosignals

8.3.1 Sensing

All devices in use today allow setting of modes with inhibited "on demand" activity. Thus, in most pacing modes, the device must sense the intrinsic cardiac activity in individual heart chambers – efficient sensing is essential for its operation. Timing intervals (such as blanking and refractory periods) also help with proper sensing. During a blanking period, sensing is completely inhibited; in refractory periods, sensed events have no effect on the timing of pacing.

According to the lead polarity, sensing may be conducted as bipolar or unipolar. Various pacing and sensing configurations may be combined. In certain types of devices, the sensitivity is also changed during a timing cycle in the bipolar configuration to restrict T wave sensing. The set sensing value is also applied in defibrillators for tachycardia detection. Pacemaker detection circuits must have suitable filter frequency characteristics so that sensing of electrical processes in the body (other than cardiac activity), in particular myopotentials, is prevented. Frequency and amplitude ranges of sensed biosignals are shown in Fig. 8.3.

8.3.2 Sensitivity

Sensing must be conducted at the appropriate sensitivity. Sensitivity is defined as the lowest input signal waveform amplitude at which device response is induced, that is, the escape interval is triggered (see Chap. 9). Per standards [17], the sensitivity in the atrium and the ventricle is defined as the amplitude of the standard test signal voltage, which is just sufficient to be detected by the device. Laboratory measurements of sensitivity are described in the standard. Today, the field is well covered technically; in the past, however, inhibition characteristics were quite important in terms of practice. Excessive sensitivity may cause sensing of far-field R waves, T waves, noise, myopotentials, or external electromagnetic interference; in inhibitory modes, it can lead to pacing inhibition. Insufficiently sensitive sensing may cause redundant and asynchronous or competitive pacing or pacing in the vulnerable phase.

Devices sense using lead sensing electrodes implanted in heart chambers. The sensitivity of the intracardial signal sensing is programmable, and the programmable value of the sensitivity setting represents the threshold value, determining the minimum electric amplitude sensed by the device as the intrinsic cardiac activity in the particular heart chamber. Higher programmed values mean lower sensitivity to intrinsic electric cardiac activity and vice versa.

Today it is often possible to program automatic sensitivity in each sensing channel. The device then automatically adapts the sensitivity value to the level of a sensed signal by means of automatic gain control system of the input amplifier. The sensitivity level is updated in each cardiac cycle, and the sensitivity range may differ for the atrium and the ventricle. The level is set based on the average of the measured sensed events and noise levels. For good functioning, a minimum value of intrinsic signals – for example, a P wave of 1.0 mV and an R wave of 5.0 mV – is recommended.

8.3.3 Pacing Pulse

A negative voltage pulse is needed for pacing. The amplitude and width are optional parameters of the pacing pulse. It should be noted that the total pacing energy is directly proportional to the second power of the amplitude and the first power of the pulse width according to the relationship $E = U^2/(R*T)$. During manufacturing, the parameters of the pulse amplitude and width are measured under standard conditions per the standard [17]: 37 °C, rate 60 pulses/min, amplitude 3.5 V, pulse width 0.4 ms, and nominal sensitivity at three load values.

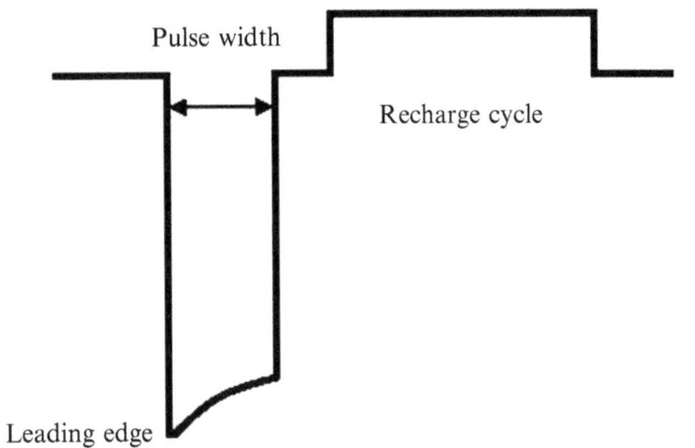

Fig. 8.4 Output pacing pulse parameters

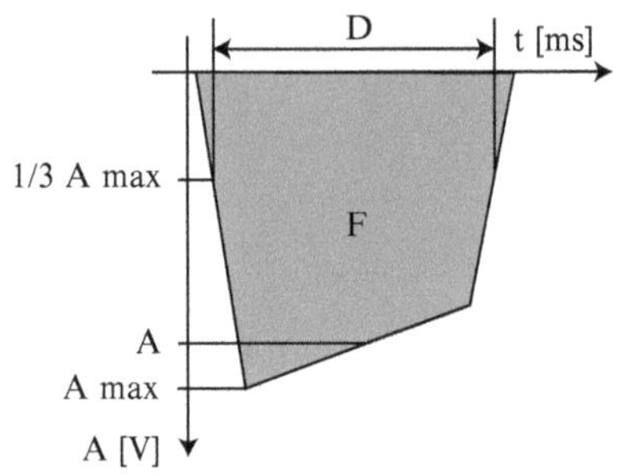

Fig. 8.5 Pulse amplitude and width measurement according to ISO 14708-2

8.3.3.1 Pulse Amplitude

The pulse amplitude or output pulse voltage is measured as the voltage of the leading edge of a device output pulse. The amplitude is an independently programmable parameter, the values of which can range from 0 (pacing is off) to 7.5 or 8.4 V or more, according to the type of device. Pacemakers have a lower output capacity to improve the output signal and are designed with circuits that recharge quickly upon the discharging of an output coupling capacitor following a paced pulse. This circuit ensures proper amplitude pulses at higher rates and improves sensing circuit recovery after an output pulse. Recharging the output circuit capacitor appears on lead pacing electrodes as a low-amplitude pulse of opposite polarity immediately after the output pacing pulse (Fig. 8.4).

Most up-to-date pacemakers are designed with a function for automatic setting of the output pulse amplitude and monitoring of the pacing efficiency. The purpose is to adapt pacing pulses dynamically so that reliable pacing is ensured and the output voltage is optimized simultaneously. The evaluation of the successfulness of pacing is based on the resulting electrogram (EGM). The EGM is sensed by either the electrodes delivering the pacing or a special auxiliary sensing electrode available in certain systems. Each algorithm adds a certain safety reserve to the measured threshold, though. If pacing is not delivered, the pacemaker applies a back-up pacing pulse with a higher amplitude.

8.3.3.2 Pulse Width

The independently programmable parameter of pulse width determines how long the output pacing amplitude will be applied between pacing electrodes. The pulse width is programmable in the range of 0.05–2.00 ms. The pulse width is, as per the standard [17], measured in one-third of the voltage of the leading edge (Fig. 8.5).

8.4 Adaptive-Rate Pacing

Pacing modes with an adaptive rate (frequency) are devised for patients suffering from chronotropic incompetence who would profit from the increased pacing rate during increased activity. In the case of chronotropic incompetence, the physiological increase of heart rate is insufficient. Various types of failures may occur. In addition to being incapable of accelerating the heart rate, it also insufficiently increases the heart rate during activity, and decelerates the heart rate too quickly after activity or various combinations.

Sensors are used to detect changes in a patient's metabolic requirements. These are divided into three classes according to the physiological level they sense. Primary sensors sense physiological factors influencing a sinus node, for example, circulating catecholamines, and the activity of the autonomous nervous system. Secondary sensors are capable of sensing physiological parameters relating to physical activity, such as minute ventilation (MV), temperature, muscle contractions, or the QT interval. Tertiary sensors sense body motion as a consequence of physical activity. Despite the number of sensors, an ideal sensor capable of simulating the behavior of the sinoatrial node is still not available. In practice, an accelerometer is the most common sensor. Upon detection of increased activity, the algorithm translates the measured level of activity into the LRL increase above the basic value. For the purpose of pacemaker timing, the algorithm thus sets a new higher LRL. If all parameters are programmed properly, the pacing increases with increasing activity and decreases if the activity returns to the resting level.

The maximum value to which the pacing rate may increase as a consequence of sensor driving is determined by a parameter that is named differently by different manufacturers: a maximum sensor rate (MSR) or an upper sensor rate, or, in the time domain, generally a sensor-driven interval. This parameter is programmable, but usually not completely independently. The parameters of adaptive-rate pacing are related

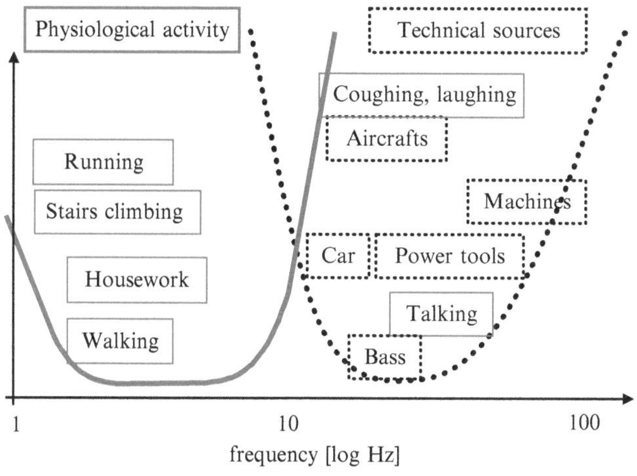

Fig. 8.6 Frequency range associated with physiological activity and technical sources

to the parameter of a maximum tracking rate and refractory periods. Adaptive-rate pacing can be unsuitable for patients suffering from angina pectoris or other symptoms of cardiac ischemia at higher pacing rates. The appropriate MSR value should be selected based on the estimation of the maximum pacing rate a patient tolerates well. A long programmed refractory period, together with a high MSR, may result in asynchronous pacing during refractory periods – the combination may lead to a small sensing window or none at all [20, 32, 63]. Because of a proven potentially proarrhythmic effect of adaptive-rate pacing, it must be programmed with caution, and the effect of higher pacing rates on the patient must be evaluated before the patient leaves the medical center.

8.4.1 Accelerometer

The accelerometer sensor detects body motion associated with a patient's physical activity. The sensor generates an electronic signal proportional to the magnitude of motion. Based on the accelerometer output signal, the pacemaker estimates the increased needs of the patient relating to the exercise and translates the signal into a pacing rate increase. The accelerometer is often placed in a hybrid integrated circuit and reacts to the activity in the frequency range typical of motion activity, that is, approximately 1–10 Hz (Fig. 8.6). The sensor evaluates the signal frequency and amplitude. The frequency reflects the motion frequency; the amplitude reflects the intensity of the motion. The algorithm translates measured acceleration into a rate increase over the LRL value. Because the accelerometer is not in contact with the pacemaker can, the response to static pressure applied to the pacemaker is negligible. In terms of design, the accelerometer sensor is a piezoelectric crystal. It senses various kinds of activities using the ratio of load in three dimensions of a

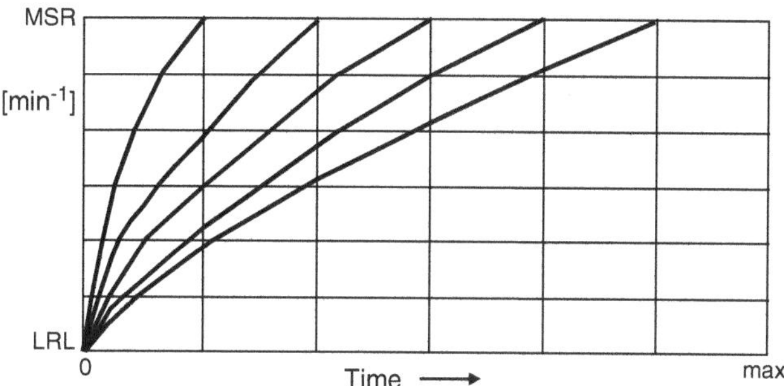

Fig. 8.7 Reaction time parameter

coordinate system, which is why the position of the implanted device has no impact on its operation.

The response of the device to the signal generated by the accelerometer sensor is determined by several programmable parameters that are named differently by different manufacturers; however, the principle is similar [32, 62, 63]. The basic parameter represents what is referred to as the activity threshold – an activity level that must be exceeded before the pacing starts to be driven by the sensor. The setting of the threshold must allow pacing during a minor activity, such as walking, to be increased but must also be high enough to prevent an inappropriate pacing increase in response to low-intensity movements (e.g., respiration, heartbeat, or Parkinson's disease–related tremor).

The response time is another parameter; it sets the time required to accelerate the pacing to the new level if an increase in activity level is detected (Fig. 8.7). The selected response time determines the time over which the pacing will increase from the LRL to MSR at the maximum activity level. A lower value of the response time will cause fast acceleration of pacing; a larger value will cause slower acceleration of pacing, or else, according to system manufacturer.

The response factor parameter is defined as the change of intervals between pacing pulses caused by the change of the accelerometer signal (Fig. 8.8). The relation between the accelerometer signal and this interval is linear. The accelerometer signal is zero at rest. The pacing is then provided at the LRL. Under maximum load, the accelerometer signal has the highest amplitude, and the pacemaker should pace at the programmed MSR. Setting the response factor to a higher value will enable the rate to reach the MSR at a lower level of activity. At a lower value, more activity is needed to reach the MSR, or else, according to system manufacturer.

The recovery time parameter determines the time required for the pacing to decelerate from the MSR to the LRL in the absence of activity (Fig. 8.9). A lower value of deceleration will cause a faster or slower decrease in the pacing rate after the decrease or cessation of the patient's activity, or else, according to system manufacturer.

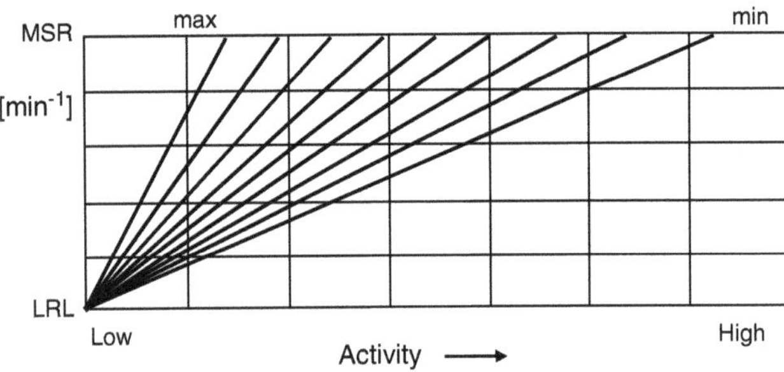

Fig. 8.8 Response factor parameter

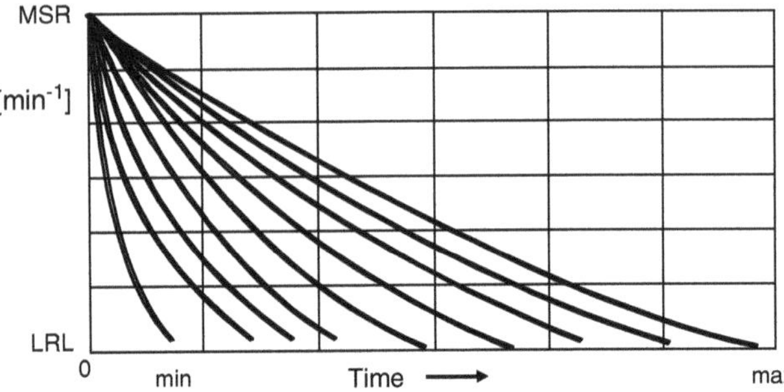

Fig. 8.9 Recovery time parameter

The effect of individual adaptive-rate parameters and their specific settings may be estimated rather relatively, based on their change from the nominal values of the devices. These were defined as optimal in clinical studies, and their good applicability can be expected in most patients. Today's devices also have algorithms for setting an automatic pacing profile.

8.4.2 Minute Ventilation

The use of respiration properties is another possibility of cardiac rhythm management. MV is a product of respiration frequency (breaths per minute) and tidal volume. The heart rate is linearly related to the MV up to the anaerobic threshold. At exercise levels beyond the anaerobic threshold, the relation is still approximately linear, but at a reduced slope. The relationship between both slopes is different in individual patients and depends on various factors such as sex, age, and exercise frequency and intensity. An MV sensor for cardiac rhythm management measures transthoracic impedance. During inspiration, the transthoracic impedance is high; during expiration, it is low. To measure this impedance, a certain manufacturer's system applies a measuring current pulse every 50 ms between the pacemaker can and the lead proximal electrode. This pulse is a flat signal with a low amplitude that does not distort the surface ECG, even though excitation waveforms can be detected and depicted in certain ECG devices. During the flow of current between the electrode

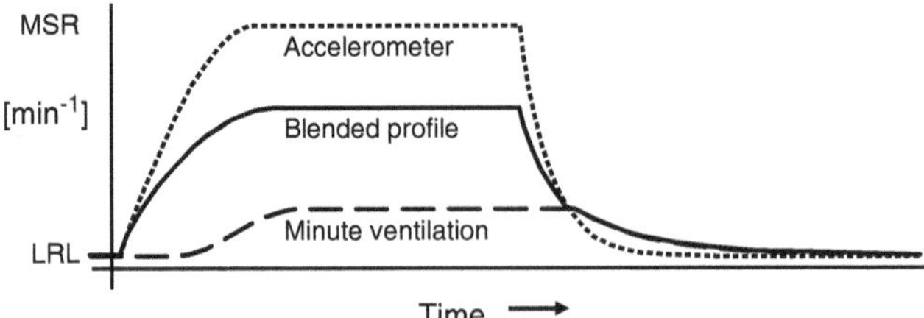

Fig. 8.10 Blended pacing profile

and the device can, an electrical field modulated by breathing is created across the thorax. The device detects the voltage between the distal electrode and indifferent electrode placed on the pacemaker header. The respiration curve is then processed to measure the total volume.

For the activation of the MV sensor, the system needs to measure the basic level of MV at rest. Increase of MV above the basic level value, caused by increased metabolic needs, will be detected by the pacemaker, and the pacemaker will translate it into a pacing rate increase using an algorithm. The relation between the detected increase in the MV and final acceleration of pacing is determined by the response factor parameter. Pacemakers allow the slope to be programmed even beyond the anaerobic threshold, simulating the physiological relationship between MV and heart rate. This parameter – the response factor beyond the anaerobic threshold – is determined as a percentage of the normal response factor.

8.4.3 QT Interval

As a consequence of accelerated heart rate and the impact of catecholamine, in particular noradrenaline, the QT interval is shortened proportionally. However, the QT interval of a paced cardiac cycle is also shortened. Therefore it is necessary to pace the ventricle to make it possible for the device to measure the QT interval, and the occurring QRS-T interval can serve for pacing rate management. The subsequent electrical response is filtered. This metabolic sensor increases the pacing proportionally according to the noradrenaline level. It can thus respond even to mental activity and changes in the body's position. Nevertheless, considerably delayed onset of response, long recovery, merely ventricular use, and relating permanent pacing [64] can be seen as disadvantages.

8.4.4 Sensor Combination

Some systems allow two sensors – the accelerometer and the MV or the accelerometer and the QT interval – to be combined. If two sensors are selected for adaptive-rate pacing, the signals from both sensors are combined, and an average pacing profile is produced (Fig. 8.10).

8.5 Hysteresis

Pacing hysteresis is an feature allowing the patient's intrinsic rhythm below the programmed LRL to be supported by temporarily prolonging the escape interval; the intrinsic heart rate is thus given preference. As a consequence, the hysteresis can extend the longevity of the device by decreasing the number of pacing pulses and optimize the patient's hemodynamics by supporting the intrinsic rhythm. Hysteresis is applied if the intrinsic heart rate is between the LRL and the LRL decremented by the set offset value.

In single-chamber pacing modes, hysteresis is available in VVI or AAI modes. The device delivers pacing at the programmed LRL. After a sensed event, the device prolongs the escape interval by the hysteresis interval. Programmed hysteresis then determines the lowest heart rate to occur before the pacing commences. Hysteresis is activated by one nonrefractory sensed event.

In dual-chamber devices, the function is available in DDD, VDD, and DDI modes. Here, hysteresis is usually activated by one nonrefractory sensed atrial event and is deactivated by an atrial or ventricular pacing pulse [65, 66].

Upon the activation of the search hysteresis function, the device prolongs the escape interval periodically by a programmed hysteresis offset value to detect possible intrinsic cardiac activity below the LRL. During the search hysteresis, the pacing decreases by the offset value over several cardiac cycles. When the search ends, hysteresis remains active if intrinsic activity is sensed during this interval. Failing that, pacing is restored at the LRL. If intrinsic cardiac activity occurs, the function allows the pacing to be inhibited until the LRL value decreased by the hysteresis offset value is reached.

8.6 Diagnostic Features of Pacemakers

Storing diagnostic information in the pacemaker helps monitor the patient's condition between follow-ups and on a long-term basis. The diagnostics also suggest which pacemaker parameters are to be adjusted to provide optimum treatment. The pacemaker records various information relating to the patient's heart rhythm and some device functions. The information is either recorded automatically or certain functions must be activated. Using the programmer, the recorded information may be downloaded, displayed, and printed.

The data normally recorded within the diagnostic functions are as follows:

- Results of automatic threshold measurement
- Current measure of P wave and R wave amplitude
- Current measure of lead impedance
- Paced and sensed event counters
- Sensor data
- Heart rhythm histograms
- Real-time intracardial EGMs
- Annotated event markers
- Automatically stored EGMs
- Patient-triggered stored EGMs
- Daily P and R wave amplitude measurements and trend
- Daily A and V lead impedance measurements and trend
- Condition of atrioventricular conduction
- Search histogram of intrinsic AV conduction
- Sensor-indicated pacing profile
- Tachycardia episodes
- Bradycardia episodes
- Atrial arrhythmia trend
- Duration of atrial arrhythmias
- Ventricular rhythm during atrial arrhythmias
- Change history of the most important parameters

Certain types of data are deleted during an ambulatory device follow-up, while others are not. With the exception of lead replacement, long-term data concerning the lead condition should not be deleted. Collection of data stops at the end of the device's longevity. Nevertheless, the collected data can be downloaded even afterward for a certain period of time [67, 68].

8.6.1 Arrhythmia Records

Data concerning tachycardia in both heart chambers include the date and time of an episode, the maximum and average rate, and the total duration of the episode. Moreover, upon the detection of an episode, a sensor-indicated pacing rate can be stated in printed reports. A programmed method of data collection determines whether episodes are updated and older episodes are rewritten, or whether collection is terminated once a set limit is reached. When displaying recorded arrhythmias, the intracardial signal time base can be adjusted; it can be rolled right or left, and electronic rulers can be used for measuring on the display.

In addition to short atrial and ventricular arrhythmia episodes, atrial tachycardia also triggers daily recording of their duration. On a long-term basis, the data allow the trend of this arrhythmia type to be identified as increasing or decreasing. The ventricular rhythm during atrial arrhythmia is displayed in the profile of a ventricular action recorded during atrial arrhythmias sensed since the last patient's follow-up. Based on the information, the extent of the ventricular pacing during atrial arrhythmias can be identified.

Arrhythmia recording can be triggered automatically (at a set triggering heart rate) or by the patient. A patient may commence recording data on EGMs, intervals, and annotated markers during a symptomatic episode by placing a magnet on the device. If the function is activated, the device stores

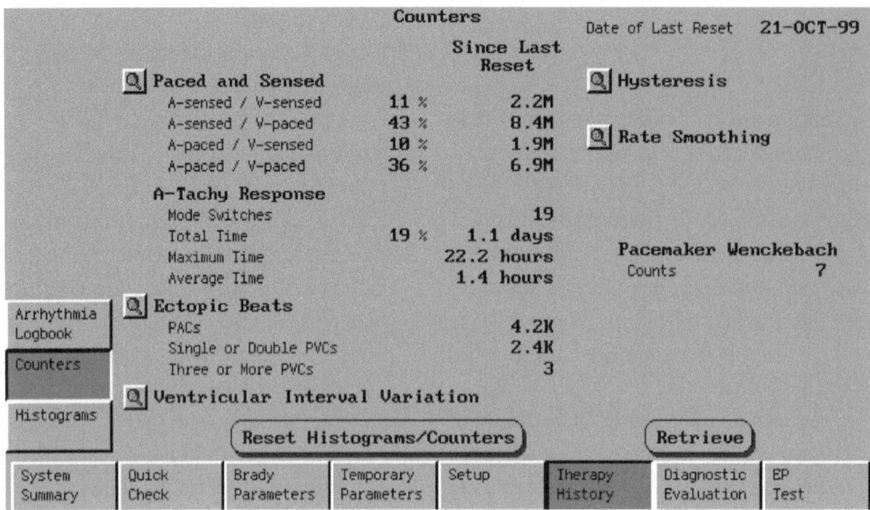

Fig. 8.11 Counters screen view [32] (© 2012 Boston Scientific Corporation or its affiliates. All rights reserved. Used with permission of Boston Scientific Corporation)

data from a period of time even before the onset and after the end of the episode in the memory. The stored data include an episode number, atrial and ventricular action at the time of magnet application, and the time and date of the magnet application.

8.6.2 Counters

Event counters record the number of intrinsic and paced events occurring during the recording in all heart chambers with active leads (Fig. 8.11). This period starts with resetting the counters and ends when the data is retrieved from the pacemaker via telemetry. Neither data retrieval nor changing the programmed parameters or the pacing mode resets the counters. The following events are mostly counted and recorded if programmed in the permanent pacing mode:

- Paced and sensed events
- Atrial tachy response (ATR) mode switch data
- Total ATR mode switches
- Total ATR mode switch time
- Maximum mode switch time
- Average mode switch time
- Atrial extrasystoles
- Ventricular extrasystoles
- Atrial tachycardia detection
- Ventricular tachycardia detection
- Ventricular interval variation
- Hysteresis
- Atrioventricular conduction hysteresis
- Sensor data

8.6.3 Histograms

The histogram function allows atrial and ventricular paced and sensed events to be recorded during a recording period, which is similar to that of counters, and to be displayed graphically (Fig. 8.12). Histograms show the range of the patient's heart rate and facilitate the evaluation of set adaptive pacing parameters in case of chronotropic incompetence. In certain systems, data collecting may be set so that data including or excluding the events sensed in what is referred to as a refractory period (see Chap. 9) are recorded.

Atrioventricular histograms show atrial sensed events followed by ventricular sensed or paced events and atrial paced events followed by ventricular sensed or paced events. Hence, the condition of atrioventricular conduction is monitored. These events are classified according to heart rate. Histograms of pacing/sensing show atrial paced and sensed events and ventricular paced and sensed events, also classified according to the heart rate. Histograms are reset together with the counters.

8.7 Electrophysiologic Testing Using a Pacemaker

Electrophysiologic (EP) testing available in certain pacemakers allows for the measurement and evaluation of atrioventricular conduction properties or noninvasive induction or termination of tachycardia. Real-time recording of EGMs, ECG curves, and event markers are enabled at the same time. The EP test allows a series of timed pacing pulses or bursts

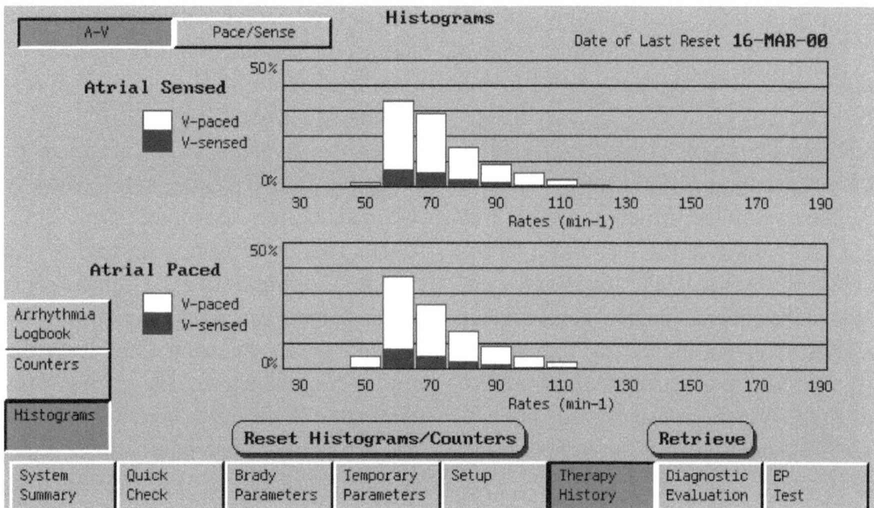

Fig. 8.12 Histograms screen view [32] (© 2012 Boston Scientific Corporation or its affiliates. All rights reserved. Used with permission of Boston Scientific Corporation)

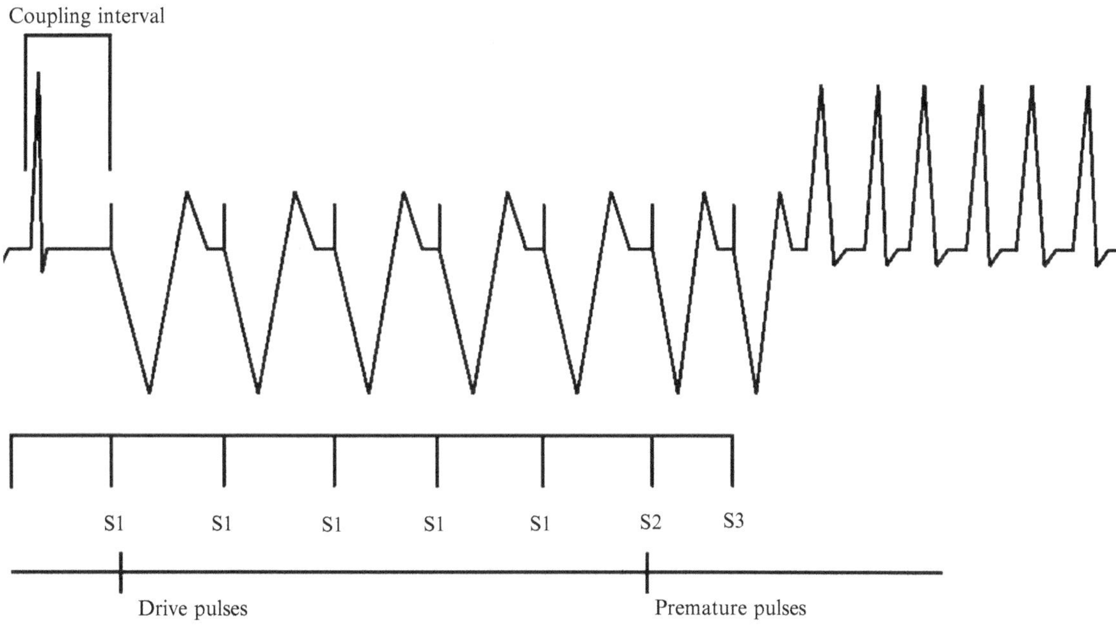

Fig. 8.13 Electrophysiologic pacing impulses [32] (© 2012 Boston Scientific Corporation or its affiliates. All rights reserved. Used with permission of Boston Scientific Corporation)

to be delivered. These pulses may be delivered to the atrium or the ventricle. In dual-chamber modes, the choice of a heart chamber is nominally set to the atrium. If the ventricle was selected, life-threatening ventricular arrhythmias might be induced. During atrial pacing, if ventricular pacing is activated in the permanently programmed normal bradycardia therapy mode, back-up pacing in the VVI mode is available. The parameters of the back-up VVI pacing may also be programmable independently from the permanent parameters. Programmed electrical stimulation allows the pacemaker to deliver a series of equally timed pacing pulses (S1) followed by premature pacing pulses (S2–Sx) to induce or terminate arrhythmia. Drive pulses (or S1 pulses) serve for pacing and cardiac rhythm management at a rate slightly higher than the intrinsic action. All pulses are delivered in the V00 or A00 mode (depending on the selected chamber), with programmed parameters of the pulse width and amplitude (Fig. 8.13).

Burst protocol allows a manual asynchronous burst in the selected chamber to be set, with an optional delay period with or without a pacing pulse and optional asynchronous back-up ventricular pacing during atrial pacing. In dual-chamber pacemakers, back-up pacing in the V00 mode is available. Otherwise, the ventricular pacing is turned off during atrial pacing and the inhibited delay period. The initial

S1 pulse or burst are synchronized by the coupling interval with the last sensed or paced event. If the permanent mode is dual-chamber, the initial interval is synchronized with a ventricular event.

EP tests consume extensive amounts of energy. Triggering a burst at a lower battery capacity may lead to a considerable temporary decrease in battery voltage, which is why it is not recommended to set pacing pulses at the maximum amplitude. When applying the EP protocol, the telemetry wand must always be kept above the patient's pacemaker. Provided that the telemetry wand is removed, the EP test may be terminated prematurely, and the permanently programmed mode can be restored. The EP functions usually are not available immediately after implantation or if an adaptive-rate mode is the main pacing mode.

The application of a pacing pulse or a burst is secured, and triggering must be reconfirmed. This function may be used only by people trained in carrying out EP testing. Pacing options of this function may be applied only if the patient is monitored carefully and an external defibrillator is prepared for immediate use.

Per the standard [17], integrated protection against an excessive output pacing rate (runaway) is required in all pacemakers; the maximum permissible pacing rate must be stated. This protection is devised to eliminate pacing acceleration resulting from most failures caused by the breakdown of one component. This internal function is not programmable and works independently from the pacemaker's main timing circuit securing the pacing. Protective circuits must eliminate the pacing acceleration over a certain value (e.g., 210 pulses/min), depending on the type of device and manufacturer. During EP testing provided by the device and related temporary pacing at a high rate, the runaway protection is temporarily deactivated.

9 Pacemaker Timing

Coordinating the timing of individual electrical processes in the heart is required for proper electrical activation. Pacemakers used for bradycardia treatment must respect this timing. The more perfectly they try to approximate natural cardiac activation, the more complex is the timing of individual processes. To comprehend how pacemakers work, it is absolutely essential to become familiar with the timing of individual events.

The behavior of pacemakers in time is, in certain contexts, expressed in the frequency domain or by the number of a certain type of event per minute (in units [Hertz, per minute, minute^{-1}]); other times it is expressed in time intervals using seconds or milliseconds and minute^{-1}) units can be easily converted using a constant of 60,000, corresponding to the number of milliseconds in 1 min. Than the equation is *quantity (ms) = 60,000/quantity (/min, min*$^{-1}$*)*. Both of these methods of describing an event are often interchanged, so at a pacing rate of 60/min, the interval between stimuli is 1 s, or 1,000 ms. In practice, both approaches are combined when programming the systems. In the time domain, refractory and blanking periods and other rather technical parameters are set. In the frequency domain, clinical parameters, such as pacing rate, intrinsic heart activity, and limits of tachycardia detection zones, are set.

For good orientation in this chapter, knowledge of North American Society of Pacing and Electrophysiology/British Pacing and Electrophysiology Group code (see Table 5.1) and single-chamber and dual-chamber pacing modes (described in Chap. 5) is required. Time relationships between sensed and paced events in the most frequently used single- and dual-chamber modes is described here. The manner of a pacemaker's response to a sensed or paced event in a particular mode depends on the manufacturer; various types of devices from the same manufacturer may also differ.

Timing of a cardiac cycle is based on the pacing lower rate interval (LRI). Sometimes, it is referred to as an automatic interval. In terms of rate, the LRI parameter corresponds to the lower rate limit (LRL). It is a minimum atrial and/or ventricular pacing rate. Asynchronous modes pace at the LRL permanently; inhibited modes occur only in the case of an absence of sensed intrinsic cardiac activity or without sensor-driven pacing. In the case of inhibited modes, that is, when intrinsic cardiac activity is sensed, the timing of single- and dual-chamber pacing systems is supplemented with an escape interval (EI). The EI is the period of time between a sensed or paced event and a possible following pacing pulse in the same heart chamber, unless intrinsic heart activity is sensed (Fig. 9.1). The origin of the term is not clear. It refers to the interval in which one activation center may escape from the dominance (control) of another center. So, for example, in a pacemaker in the VVI mode with an EI of 1,000 ms, the intrinsic cardiac activity may escape from the rhythm dominance (control) when it occurs in a period not longer than 1,000 ms since the last ventricular event. Technically, it is the period of time after the lapse of a timed event, that is, atrial or ventricular pacing. It is the time interval after which a running counter of the electronic system, set to the initial value and counting with certain clock frequency, is reset. The counter reset to the initial value by a sensed event is the basic principle of timing of inhibited pacing modes.

The EI measured by the pacemaker from the moment an event is sensed equals the LRI. Nevertheless, the EI measured on a surface electrocardiogram (ECG) from the beginning of the QRS complex cannot be determined accurately. That is why the EI seems to be a little longer than the LRI when measured manually – the device senses the intrinsic cardiac activity later than it appears on the surface ECG [69].

9.1 Timing in Single-Chamber Modes

Single-chamber pacing modes are used in patients with paroxysmal asystoles or without asystoles and in patients with chronic supraventricular arrhythmias for whom dual-chamber pacing cannot be utilized. The single-chamber pacing modes serve for pacing only in the atrium or only in the ventricle [69].

D. Korpas, *Implantable Cardiac Devices Technology*, DOI 10.1007/978-1-4614-6907-0_9,

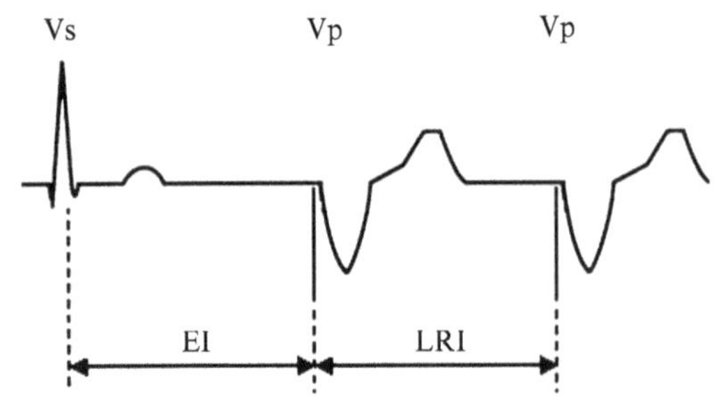

Fig. 9.1 Escape interval (EI) + lower rate interval (LRI)

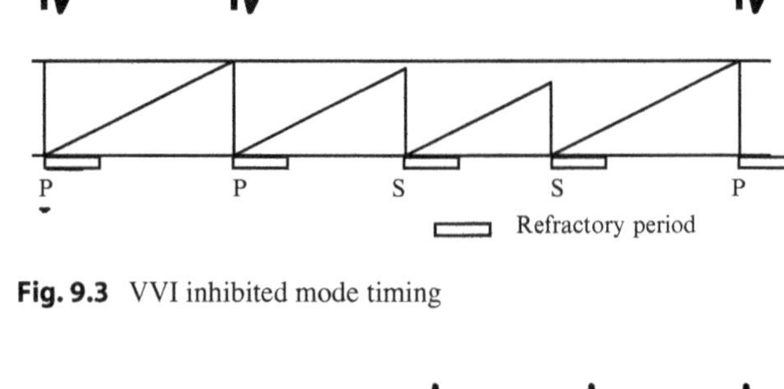

Fig. 9.3 VVI inhibited mode timing

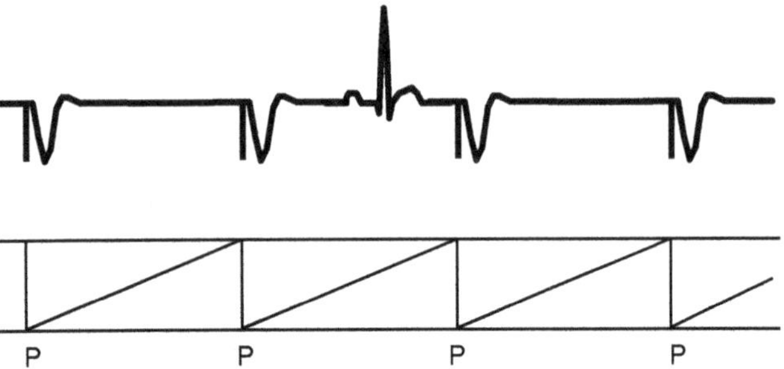

Fig. 9.2 V00 asynchronous mode timing

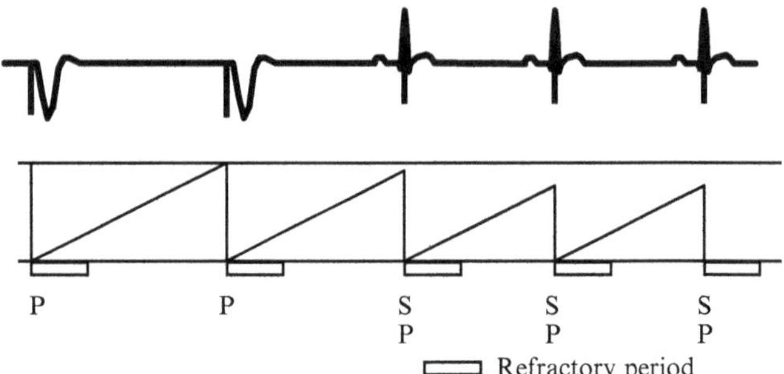

Fig. 9.4 VVT triggered mode timing

9.1.1 Asynchronous Modes

Single-chamber asynchronous modes include A00 and V00. They are the simplest pacing modes. The A00 mode provides atrial pacing at a programmed LRL without inhibition by sensed atrial events. The V00 mode provides ventricular pacing at a programmed LRL without inhibition by sensed ventricular events. Hence, sensing is not active in this mode, and no further timing interval is defined. The timing is simple: the counter runs according to the set LRL without the possibility of resetting. After the counter stops, pacing is delivered regardless of possible intrinsic cardiac activity.

The V00 asynchronous mode (Fig. 9.2) was historically the first pacing mode. Its main disadvantage (besides energy consumption) is the danger of competitive pacing, that is, the collision of a pacing pulse and intrinsic cardiac activity. Pacing timing in a vulnerable phase (T wave) is capable of inducing ventricular tachycardia.

9.1.2 Inhibited Modes

During inhibited modes (AAI and VVI), sensed intrinsic cardiac activity resets the counter and retriggers the EI. In inhibited modes, the LRL is supplemented with additional time intervals. It is a blanking period of the sensing channel and a refractory period (Fig. 9.3), described further in greater detail in Sect. 9.4. In the VVI mode, the ventricle is paced in case no intrinsic ventricular events are sensed; the setting of the ventricular refractory period avoids the inhibition of ventricular pacing by, for example, noise. Pacing is delivered after the lapse of the EI at the programmed LRL in the VVI mode or at a rate indicated by a sensor in the VVI(R) mode. An event sensed during a ventricular refractory period is classified as refractory and does not inhibit ventricular pacing.

In AAI(R) modes, the timing is identical, but with an atrial refractory period. The difference lies in the sensing sensitivity, which must be higher for atrial sensing (i.e., a lower value in millivolts).

9.1.3 Triggered Modes

In single-chamber triggered modes, including AAT and VVT modes, sensed intrinsic cardiac activity triggers pacing and, at the same time, resets the counter. These modes are rarely used. In the VVT mode, pacing is delivered at the LRL. If an intrinsic R wave is sensed before the lapse of EI, a pacing pulse is applied immediately in the ventricle (Fig. 9.4). The pacing pulse will appear on the ECG somewhere during the intrinsic QRS complex. Basically, AAT works on the same principle – the difference is that it senses a P wave and paces in the atrium. Upon the absence of intrinsic rhythm, the modes are identical to VVI or V00. Naturally, the triggered modes may sense even external interference as intrinsic cardiac activity.

The VVT mode may be used for diagnostic purposes. The upper pacing rate (also upper rate or maximum pacing rate [MPR]) is limited by the refractory period and runaway protection [94].

9.2 Timing in Dual-Chamber Modes

Dual-chamber modes are much more complex than the above-described single-chamber modes. Many more timing parameters are added, among which numerous interactions exist [94].

The atrioventricular (AV) delay parameter (AV delay; AV interval [AVI]) is a programmable time interval from the occurrence of a paced or sensed event in the atrium until the occurrence of a paced event in the ventricle. AV delay sustains AV synchronization of the heart. The AV delay parameter is used in all dual-chamber modes. In VDD, DDI, and DDD modes the pacemaker delivers ventricular pacing at the end of the AVI if the pacing is not inhibited by spontaneous ventricular depolarization or if the maximum tracking rate (MTR) is not exceeded. In the D00 mode the device delivers ventricular pacing at the end of the AVI regardless of spontaneous ventricular depolarization. The delay between a sensed atrial event and corresponding ventricular pacing represents the AV delay after a sensed event (interval atrial sensing – ventricular pacing = sensed AVI [SAV]). The delay between a paced atrial event and corresponding ventricular pacing represents the AV delay after a paced event (interval atrial pacing – ventricular pacing = paced AVI [PAV]) (Fig. 9.5).

If an EI ends sooner than an atrial event is sensed, the device will pace the atrium and then plan ventricular pacing to be delivered at the end of the PAV. The SAV can be programmed to a value lower than or equal to the PAV value. The influence of the SAV parameter on hemodynamics depends on how appropriate mutual timing of atrial and ventricular contractions is. Atrial pacing triggers atrial contraction, while sensing is conducted only during the contraction. Hence, if the SAV is programmed to the same value as the PAV, the hemodynamic AVI will differ for paced and sensed atrial events. If a ventricular event is sensed during the SAV or PAV, ventricular pacing is inhibited. The SAV is programmed to be 30–60 ms shorter than the PAV. The hidden delay between the actual cardiac event in the atrium and the time of its detection by the device is thus compensated for. The highest cardiac rate at which all sensed atrial events are tracked to ventricular pacing is referred to as the MTR or upper tracking rate parameter.

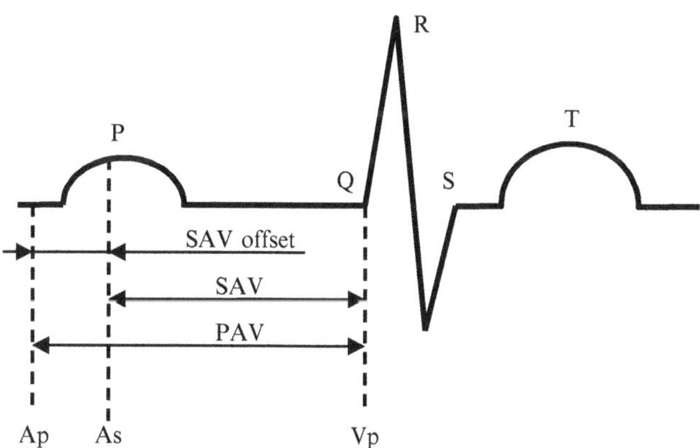

Fig. 9.5 Atrioventricular delay periods

9.2.1 D00 and 0D0 Modes

In theory, even D00 and 0D0 modes can be classified as dual-chamber modes. The D00 mode provides AV sequential asynchronous pacing at a programmed lower frequency and PAV delay, without the possibility of inhibition by sensed cardiac activity. Hence, in this mode, sensing is not conducted in any chamber. In contrast, in the 0D0 mode, rhythm is sensed in both chambers without any pacing. This mode can be temporarily programmed for diagnostic purposes; permanent programming must be avoided for safety reasons.

9.2.2 DVI Mode

The DVI mode provides dual-chamber AV sequential pacing without atrial sensing. Because the intrinsic atrial activity cannot be tracked to the ventricle, the mode is not regarded as a tracking mode. The atrial pacing is delivered at the LRL, with subsequent ventricular pacing after the lapse of PAV if it is not inhibited by intrinsic ventricular action. Sensed nonrefractory ventricular events during a VA interval trigger a new VA interval. This mode was commonly used before the DDD mode is introduced. The mode is not in use any more, and it has been left as a programmable option for diagnostic purposes or if loss of sensing occurs in the atrial channel in the DDD mode [94].

9.2.3 DDI Mode

The DDI mode delivers dual-chamber AV sequential pacing with atrial sensing, but without tracking the atrial activity to ventricles. So, it is a nontracking mode, like the DVI mode. Atrial pacing is delivered at the LRL, with subsequent ventricular pacing after PAV if an intrinsic R wave is not sensed. In DDI(R) modes, sensed atrial events are not tracked. P waves sensed in the atrium inhibit atrial pacing; the AVI is not triggered, though. The ventricle is paced at given pacing rate, and it is not a tracking mode. Ventricular pacing during sensing of the atrial event is timed according to the LRL. Sensed nonrefractory ventricular events during the VA interval trigger a new VA interval. If the intrinsic atrial rate exceeds the LRL, the atrial channel becomes inhibited, and only the VVI mode is applied in the ventricle.

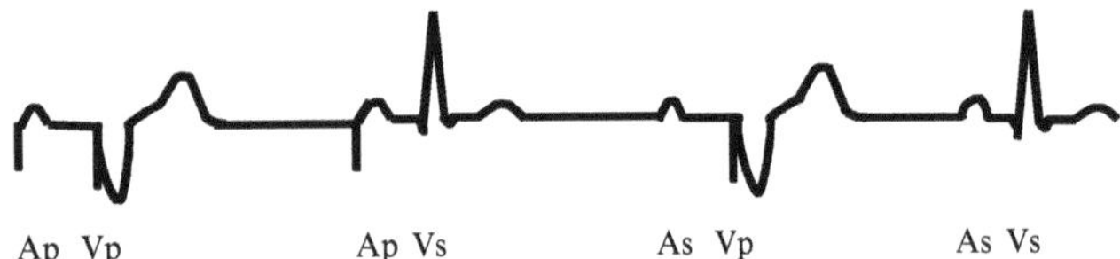

Fig. 9.6 Four basic cycles in DDD mode

9.2.4 VDD Mode

This mode delivers the ventricular pacing triggered by the atria up to the MTR. Sensing is conducted in both the atrium and in the ventricle, whereas pacing is delivered only in the ventricle. To ensure synchronized pacing at low rates, sensed atrial activity at the end of the LRI is followed by maximum SAV. As a result, the ventricular LRL is prolonged. Sensed nonrefractory ventricular events during a VV interval not preceded by sensed atrial events can be evaluated as a premature ventricular contraction and trigger a new VV interval. The advantage of the VDD mode lies in the possibility of inserting only one lead into the ventricle, with an auxiliary sensing electrode in the atrium. However, atrial pacing is impossible when a low intrinsic rate occurs.

9.2.5 DDD Mode

The DDD mode is referred to as a universal pacing mode. It provides dual-chamber pacing, dual-chamber sensing, and dual response to sensing (inhibiting or triggering). Triggering occurs if a sensed atrial event triggers an AVI, which is followed by pacing in the ventricle. This process, mentioned earlier, is referred to as tracking and sustains the synchronization of the atrium and ventricle [94].

In principle, four combinations of atrial (A) and ventricular (V), sensed (s) and paced (p) events are possible. They constitute four basic timing cycles of dual-chamber modes [69], shown in Fig. 9.6. Timing intervals in the DDD mode may vary in each cardiac cycle according to the patient's intrinsic heart rate and the condition of AV conduction, and they make use of either PAV or SAV [94].

- Atrial and ventricular pacing (Ap–Vp): delivered if the patient's intrinsic atrial rate is slower than the LR, and the intrinsic AV conduction is either damaged (AV block) or slower than the programmed AV delay.
- Atrial pacing, ventricular sensing (Ap–Vs): delivered if the patient's intrinsic atrial rate is slower than the LRL and the intrinsic AV conduction is normal, followed by an intrinsic QRS complex before the lapse of AV delay.
- Atrial sensing, ventricular pacing (As–Vp): delivered if the patient's intrinsic atrial rate is faster than the LRL and the intrinsic AV conduction is either damaged (AV block) or slower than programmed AV delay.
- Total inhibition (As–Vs): occurs if the patient's intrinsic atrial rate is faster than the LRL and the intrinsic AV conduction is normal, followed by an intrinsic QRS complex before the lapse of AV delay.

9.3 Blanking Periods

Paced and sensed events are followed by blanking periods. Blanking is the initial portion of the refractory period in which sensed signal input amplifiers are deactivated; the pacemaker completely ignores the events in the blanking period. This makes a difference if compared with refractory periods, described later, in which the device records events but they have no direct impact on its operation. Blanking thus keeps the pacemaker from sensing and misinterpreting intracardial signals and pacing artifacts. Blanking periods are triggered by sensed or paced events. They prevent far-field sensing (crosstalk) and, as a consequence, inhibit undesirable pacing; they also prevent sensing of intrinsic pacing pulses, depolarization after pacing, T waves, or excessive sensing of the same event. Immediately after a sensed or paced atrial or ventricular event, sensing is blanked in the particular chamber, mostly by a nonprogrammable period from 50 to 100 ms.

The actual time of the blanking period can be set dynamically by the device, depending on the intensity and duration of the sensed signal [94]. The dynamic blanking avoids repeated sensing of the same signal, thus minimizing the blanking period. During the blanking period, the sensing circuit in one chamber also ignores possible sensed electrical activity generated by a pacing pulse in another chamber; ventricular pacing sensed in the atrium would lead to inadequate ventricular pacing because the device would make an effort to sustain the AV synchronization. If, in contrast, atrial pacing was sensed in the ventricle, it would result in an inhibition of ventricular pacing and, as a consequence, an undesirable decrease in the pacing rate. For that reason, in DDD(R), DDI(R), and VDD dual-chamber modes, ventricular pacing triggers an atrial blanking period, and in DDD(R) and DDI(R) modes, atrial pacing triggers a ventricular blanking period. Blanking periods following paced events are equal to or longer than blanking periods following sensed events to prevent sensing of atrial and ventricular depolarization.

In general, the following blanking periods can be defined in dual-chamber modes to prevent sensing of events in the same chamber (Fig. 9.7):

- Atrial blanking after atrial sensing. Throughout this period, atrial sensing after a sensed atrial event is deactivated.
- Atrial blanking after atrial pacing. Throughout this period, atrial sensing after a paced atrial event is deactivated.

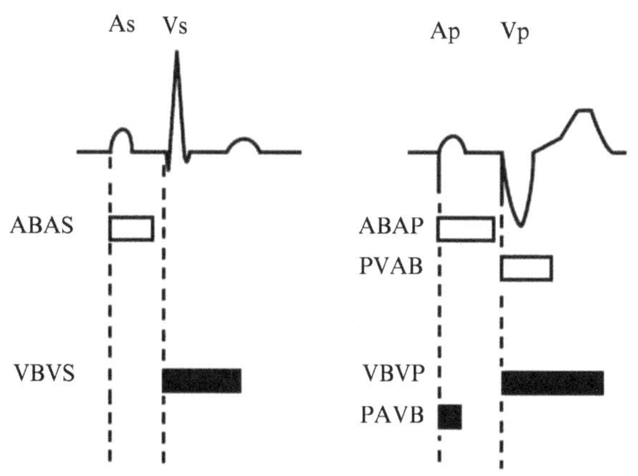

Fig. 9.7 DDD blanking periods

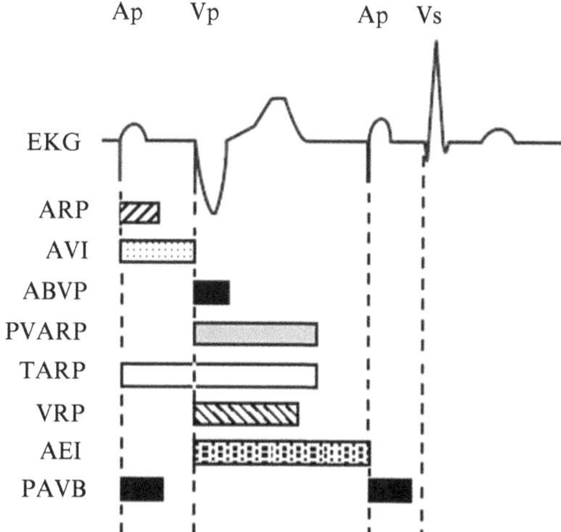

Fig. 9.8 DDD refractory periods

- Ventricular blanking after ventricular sensing. Throughout this period, ventricular sensing after a sensed ventricular event is deactivated.
- Ventricular blanking after ventricular pacing. Throughout this period, ventricular sensing after a paced ventricular event is deactivated.

All the blanking periods listed above must be made available by all manufacturers and in all types of devices [20, 63]. In dual-chamber modes, atrial pacing also triggers postatrial ventricular blanking (PAVB) in the ventricular channel. This period avoids the inhibition of the ventricular channel and subsequent asystole because of sensing of an atrial event in right ventricle. If the PAVB is too short, crosstalk may occur, which may be sensed by the ventricular channel. On the other hand, if the PAVB is too long, the pacemaker may not sense premature ventricular action (e.g., ventricular extrasystoles), which may lead to ventricular pacing during intrinsic ventricular action.

The length of a noise rejection period is in the order of tens of milliseconds, and it is made available in certain forms by all manufacturers. Sometimes it may be programmable. Any time a signal is detected during a noise detection period, the period is restarted at the moment the signal was detected. Ventricular pacing triggers the noise rejection period and atrial blanking period (postventricular atrial blanking [PVAB]) in the atrial channel. It may be programmable. The aim is to set the length of blanking so that far-field R wave sensing in the atrium is prevented and, simultaneously, the possibility of sensing spontaneous atrial activity is maintained. The length of blanking in the other chamber after a pacing pulse may differ for unipolar and bipolar sensing configurations or depending on the pacing pulse amplitude – it is usually longer for a unipolar configuration and higher amplitude. Some systems distinguish atrial blanking during ventricular pacing and atrial blanking during ventricular sensing. VA interval measurement can be used to check the far-field R wave sensing.

Certain manufacturers' systems are equipped with a sophisticated PVAB system to eliminate far-field R waves sensed in the atrium [20]. Atrial events sensed during PVAB are used to measure tachycardia.

9.4 Refractory Periods

Refractory periods are phases after a paced or sensed event when input amplifiers of the sensed signal are already activated and the device senses possible intrinsic cardiac activity (Fig. 9.8). The pacemaker timing is not, however, affected by this sensed activity. During the refractory period, a sensed event cannot trigger a particular timing interval in a particular channel. Each refractory period starts with a blanking period, during which no sensing is performed. Like the blanking periods, the purpose of refractory periods is also to exclude triggering or inhibiting the pacemakers' operation based on the detection of inappropriate signals, such as retrograde P waves, far-field R waves, or noise. Possible sensed events are classified as refractory, and the response to these events is adapted to this. The programmability of refractory periods depends on the selected pacing mode and on the manufacturer.

In single-chamber atrial modes (AAT, AAI[R]), the atrial refractory period (ARP) is defined as an interval after an atrial event, be it paced or sensed, in which sensed atrial activity does not inhibit or trigger an atrial pacing pulse.

A postventricular ARP (PVARP) is defined only in dual-chamber pacing modes (with the exception of D00). The PVARP follows a paced, sensed, or refractory sensed ventricular event, and an atrial event during this period is then classified as refractory. Hence, it does not inhibit atrial pacing or trigger an AVI after a sensed event. The PVARP avoids the inhibition of atrial pacing caused by sensed far -R waves, retrograde P waves, or noise. The PVARP must be programmed to a value longer than the time of retrograde VA conduction, so that retrograde atrial depolarization is not sensed in the atrial channel, which avoids possible occurrence of pacemaker-mediated tachycardia (PMT). If the PVARP value was shorter than the retrograde VA conduction, the probability of PMT occurring might increase. Certain devices allow the PVARP to be set automatically based on the changes in a patient's spontaneous or paced cardiac action. A sensed atrial event occurring during the PVARP is classified as refractory, does not inhibit atrial pacing, and is not tracked [32, 63]. A long PVARP shortens the atrial detection window. Programming a long PVARP in combination with a short AVI may cause a 2:1 block, occurring intermittently at a programmed MTR.

Certain systems allow dynamic PVARP and dynamic AV delay to be programmed, which optimizes the sensing window at higher pacing or sensing rates. This decreases the occurrence of 2:1 tracking or the Wenckebach behavior in DDD(R) and VDD(R) modes. A dynamic PVARP also decreases the probability of PMT at lower rates and the risk of competitive atrial pacing. The pacemaker then calculates the value of the dynamic PVARP using the weighted average of the preceding cardiac cycles. This results in linear shortening of the PVARP with increasing pacing or sensing rate. When the average cardiac rate is between the LRL and MTR, the pacemaker calculates the dynamic PVARP based on the linear dependence. If the average rate is equal to or lower than the LRL, the value of maximum PVARP is used. If the average rate is equal to or higher than the MTR interval, the value of minimum PVARP is utilized.

A ventricular refractory period (VRP) is defined as an interval after a ventricular event, be it paced or sensed, during which electrical activity sensed in the ventricle does not inhibit the pacemaker. This period is available in any mode in which ventricular sensing is activated. The use of a long VRP, however, shortens the ventricular detection window, and its programming to a value exceeding the PVARP may lead to competitive pacing because an atrial event may be sensed after the PVARP, and upon spontaneous tracking to a ventricle, it would fall into this refractory period. That being so, the pacemaker would not sense ventricular depolarization and would pace at the end of the AVI, which would result in competitive pacing.

9.5 DDD Timing Intervals

In dual-chamber modes, timing is gradually supplemented with several intervals [69]. In the ventricular channel, it is:

- LRI as the longest interval between a sensed or paced ventricular event and following ventricular pacing not inhibited by sensing.
- VRP as an interval triggered by a paced or sensed ventricular event, during which the LRI cannot be triggered again.
- AVI as an interval between a sensed (SAV) or paced (PAV) ventricular event and planned ventricular pacing. It is a substitution for the intrinsic PR interval. During this interval, the atrial channel is refractory, and a new AVI cannot be triggered.

For the atrial channel, an atrial EI (AEI) can be calculated as AEI = LRI − AVI. It is an interval between a sensed or paced ventricular event and following atrial pacing not inhibited by sensing.

The above-mentioned PVARP is a separate interval in the atrial channel; it occurs after a sensed or paced ventricular event during which an AVI cannot be retriggered by a sensed atrial event. This interval avoids inappropriate sensing of ventricular events in the atrial channel and retrograde P waves.

Total ARP (TARP) can be calculated as TARP = AVI + PVARP (Fig. 9.9).

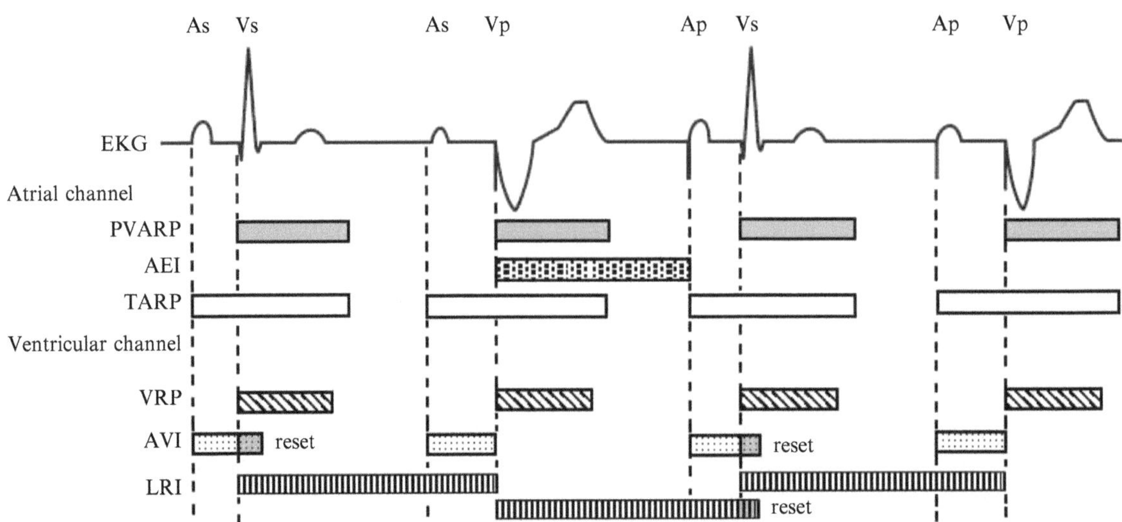

Fig. 9.9 DDD basic timing periods

9.6 Tracking Atrial Rhythm to Ventricles

In tracking modes of dual-chamber pacing, sensing and pacing in either the atrium or ventricle may occur, or in the case of biventricular pacing, possibly in both ventricles. In a tracking mode, the device responds to a sensed intrinsic atrial event by planning ventricular pacing. The delay between a sensed atrial event and corresponding ventricular pacing represents a programmed SAV. If an existing AEI is not terminated quickly by a sensed intrinsic atrial event, the device paces the atrium and then plans ventricular pacing to be delivered after a programmed PAV. If a ventricular event is sensed during the SAV or PAV, ventricular pacing is inhibited. A sensed atrial event occurring during the PVARP is classified as refractory; it does not inhibit atrial pacing and is not tracked. DDD(R) is the only mode to be considered a fully dual-chamber and tracking pacing mode. In the absence of intrinsic atrial activity, pacing in the DDD mode is delivered at a programmed LRL. In DDD(R) mode with adaptive pacing rate, pacing is delivered at a sensor-indicated rate.

9.6.1 Dynamic AV Delay

Shortening the AV delay with an accelerated cardiac action is a physiological response of the heart. The AVI may be either programmed to a fixed value or calculated dynamically based on the preceding AA interval. With a constant AVI, it is difficult to set an optimum value of the AVI to meet the patient's needs. At higher rates, a short AVI is appropriate to avoid symptomatic 2:1 block during loading and asynchronous pacing. At lower rates, a long AVI is appropriate to support the intrinsic AV conduction; hemodynamics may thus improve. So, if a constant AV delay is programmed, the AV delay value remains unchanged upon the increase of the heart rate. When using a dynamic AV delay, more physiological AV coupling is achieved in the entire range of programmed rates; the size of the sensing window is maximized at higher rates by automatic shortening of a PAV or SAV after each interval upon the atrial rate increase. Thus, the risk of Wenckebach behavior occurring is minimized in cases in which the atrial rate exceeds the MTR, the occurrence of great changes at the upper limit is minimized, and 1:1 tracking at higher rates is enabled.

Figure 9.10 shows an ECG record with dynamic AV delay. When the AA atrial interval is between the LRL and the higher of the MTR or maximum sensor rate (MSR) values, the pacemaker determines the dynamic AV delay according to the preceding AA interval based on linear dependence, as shown in Fig. 9.11. This relation is determined by programmed values of minimum AV delay, maximum AV delay, LRL, and the higher of the MRT or MSR values.

Another possibility of compensating for the time difference between a paced and spontaneous atrial event is the activation of the SAV offset function. As a result, the AV delay is shortened after a sensed atrial event by a programmed value. As a consequence, the hemodynamic AVI is different for paced and sensed atrial events. If a constant AV delay is set, the SAV offset parameter will also be fixed at a programmed value. If a dynamic AV delay is set, the pacemaker calculates the SAV offset parameter based on the intrinsic atrial rate. As a response to narrowing of the P wave in the period of increased metabolic requirements, the SAV offset is shortened linearly from a programmed value corresponding to the LRL to a value determined by the proportion of minimum AV delay and maximum AV delay and the higher of the MRT or MSR values.

9.6.2 Upper Rate Behavior

Because of the risk of induction of ventricular tachycardia by fast ventricular pacing, a dual-chamber pacemaker working in the DDD(R) mode may safely track the atrial rhythm only up to a certain rate. The upper rate behavior occurs if the patient's intrinsic atrial rhythm is faster than the MTR, exceeds the limits of atrial sensing determined by TARP, or both. It occurs only in patients with AV conduction failure; with normal AV conduction, the intrinsic atrial action is spontaneously conducted to ventricles.

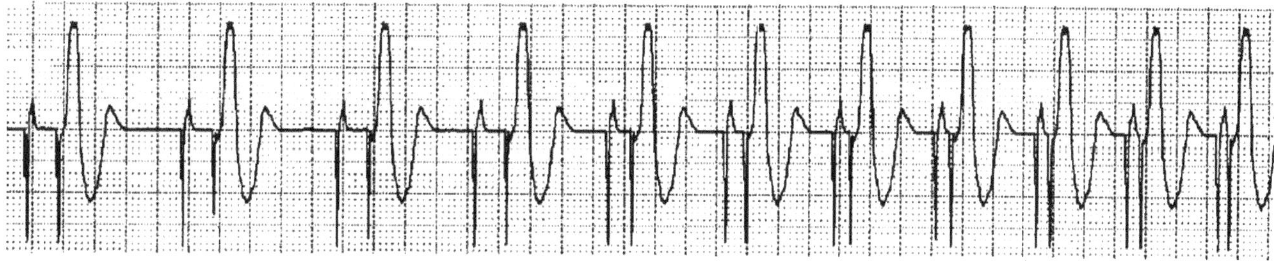

Fig. 9.10 Dynamic atrioventricular delay at a surface electrocardiogram [32] ()

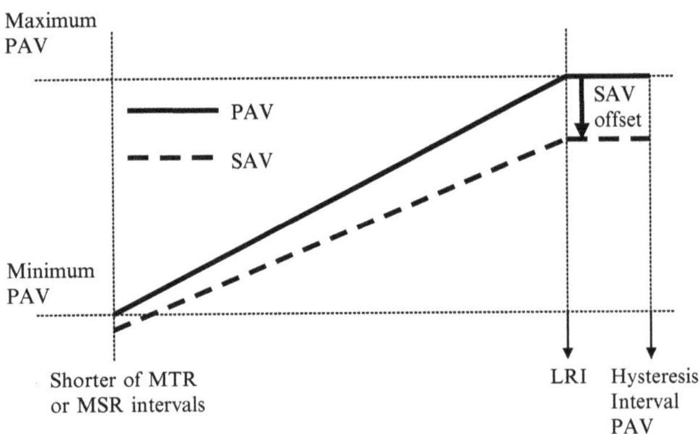

Fig. 9.11 Relationship between dynamic atrioventricular delay and the lower rate limit [32] (© 2012 Boston Scientific Corporation or its affiliates. All rights reserved. Used with permission of Boston Scientific Corporation)

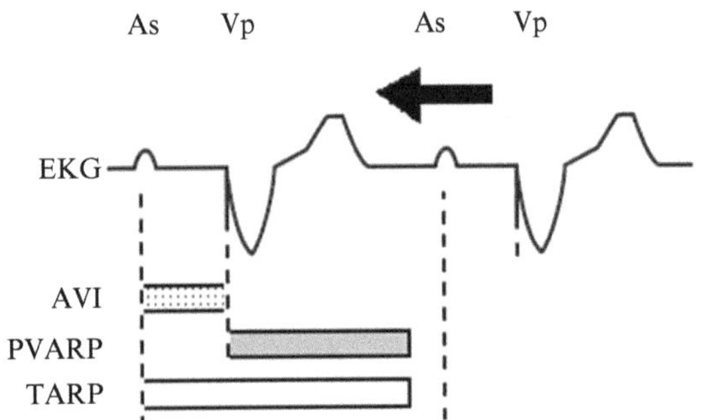

Fig. 9.12 Increasing atrial rate

If a sensed atrial rate is within the range of the programmed LRL and MTR values, ventricular pacing is delivered after the lapse of SAV or dynamic AV delay. All sensed atrial activities are tracked to corresponding ventricular pacing in the respective cardiac cycle. Nevertheless, if the atrial rate increases (Fig. 9.12), at certain rate, a P wave occurs in the PVARP.

As a consequence, loss of conduction of certain atrial events occurs as the device synchronizes ventricular pacing with the sensed P wave. If the sensed atrial rate still increases over the MTR, further atrial events will occur during PVARP. These events are not tracked and do not trigger an AVI, and the proportion of sequentially paced ventricular events to sensed atrial events will decrease. This is referred to as Wenckebach behavior. More P waves than paced QRS complexes are identified on surface ECG. Upon acceleration of the atrial rate, 5:4, 4:3, 3:2, to 2:1 blocks would gradually occur. Finally, only every second sensed atrial event is tracked; ventricular pacing is thus half the atrial rhythm. A 2:1 block can be desirable for the prevention of fast ventricular pacing at the onset of atrial tachycardia, but upon activity, that is, upon physiological increase of atrial action, it is the opposite, and the ventricular pacing may drop suddenly.

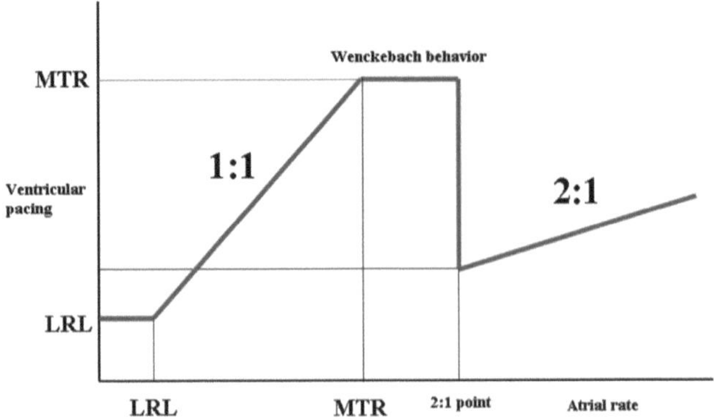

Fig. 9.13 Upper atrial rate behavior

The maximum rate at which sensed atrial events are tracked to ventricular pacing in a 1:1 ratio is determined by the MTR parameter. The MTR applies to tracking pacing modes, in particular to DDD(R) and VDD(R). The response of the device at fast atrial action in these modes is determined by the relationship between the TARP and the MTR or MSR interval. The highest atrial action the device is capable of tracking before a 2:1 block is determined by the relationship 60,000/TARP, where TARP = SAV + PVARP. The system does not allow the programmed MTR interval to be shorter than the TARP, and if the TARP is shorter than the interval corresponding to the programmed MTR, the pacemaker reduces the rate of ventricular pacing by Wenckebach behavior at the MTR value. If certain algorithms for the prolongation of the AVI are on, such behavior may occur at rates lower than MTR. Figure 9.13 shows the procedure of tracking at increasing atrial rate. The above-described pacemaker behavior is referred to as upper rate behavior or upper rate response.

9.7 Automatically Switching Modes

If a device works in the VDD(R) or DDD(R) mode, tracked atrial tachycardia may lead to fast ventricular pacing, which is why the device must be capable of ceasing to track pathologically fast sinus rhythm to ventricles during episodes of atrial tachycardia. This is made possible by an automatic mode switching on the detection of fast atrial action. The system's behavior when atrial tachycardia occurs is determined by the mode switch (MS) or atrial tachy response. If detected atrial activity exceeds a set value, the MS switches the pacing mode from a tracking mode to a nontracking mode (e.g., from DDD(R) to DDI(R) or VDI(R); from VDD(R) to VDI(R) or even VVI(R)). However, the modes with a sensor are not commonly used for MS. Until the mode switches, the ventricular pacing rate has the value of the MTR or shows signs of a 2:1 block or Wenckebach behavior. The pacing mode, to which the device is automatically switched to when the MS condition is met, is referred to as a fallback mode.

The pacemaker in the fallback mode gradually decrements the ventricular pacing to, for example, the value of the fallback LRL, a sensor-indicated rate, or an algorithm of ventricular rhythm stabilization, whichever is higher.

9.7.1 Regulation of Ventricular Rhythm

The purpose of regulation of ventricular rhythm through pacing is to attenuate the symptoms of tracked atrial fibrillation or atrial flutter by the regulation of ventricular pacing during episodes of tachycardia. The variability of the VV interval must be decreased during tracked atrial arrhythmias: very long VV intervals must be eliminated, and very short VV interval occurrence must be reduced. This option is suitable in the DDD(R) dual-chamber mode for patients with paroxysmal atrial fibrillation, and in the VVI(R) single-chamber mode for patients with permanent atrial fibrillation.

In a dual-chamber mode, the regulation of ventricular pacing is activated upon the detection of onset of an episode of atrial tachycardia. The pacemaker stabilizes ventricular pacing either by setting the pacing slightly below the level of average ventricular action, or, in another manufacturer's system, based on the weighted sum of a relevant VV cycle and preceding pacing intervals. In the former, the pacing slightly accelerates after each sensed ventricular event but never exceeds the maximum set rate. If no ventricular event is sensed, the pacemaker decelerates the pacing until another ventricular event is sensed or the LRL is reached. If the pacemaker detects the end of an episode of atrial tachycardia, the ventricular pacing regulation is turned off. In the latter, paced intervals have more influence than sensed intervals, so the paced events cause a decrease in pacing rate. This is because of the weighted-sum methodology stated above. The indicated pacing rate also depends on the LRL and MPR. As soon as a tracking mode is restored after termination of the arrhythmia, the algorithm is turned off.

If the ventricular rhythm regulation is on, it is permanently active in the VVI(R) single-chamber mode. The VVI(R) mode does not include atrial sensing; hence, permanent atrial tachycardia is assumed.

9.8 Pacemaker-Mediated Tachycardia

In DDD(R) and VDD dual-chamber modes, the pacemaker may sense retrograde conducted P waves falling outside the PVARP, which may result in triggered ventricular pacing reaching MTR. That being so, it is called PMT, also referred to as endless loop tachycardia. PMT often starts with premature ventricular contraction, which may be either intrinsic or paced. The electric impulse is conducted retrogradely through a cardiac conduction system to the atrium. If this retrograde P wave occurs after the end of a PVARP, it is sensed by the pacemaker (Fig. 9.14). This triggers an AVI, after the lapse of which the ventricle will be paced. The cycle then starts again [94].

The detection of PMT depends on the manufacturer. In one system, the PMT condition is met by counting 16 consecutive ventricular pacing pulses at a rate equal to the MRT following sensed atrial events. During these 16 intervals, the VA interval is also monitored, and it is evaluated whether PMT has occurred or whether the intrinsic atrial rate reaches or exceeds the MTR. The VA intervals are compared with the initial VA interval measured during 16 paced ventricular events. If any of the subsequent intervals is shorter or longer than the initial interval by more than 32 ms, the rhythm is declared an Wenckebach event; the counter of Wenckebach events counts the event, and the algorithm continues monitoring subsequent ventricular stimuli to detect possible PMT. If all VA intervals meet the 32-ms criterion, the rhythm is declared to be PMT. Another system defines the PMT as eight consecutive Vp–As intervals shorter than 400 ms. When the pacemaker detects the PMT, according to set rules, it prolongs the PVARP parameter in one cardiac cycle. As the following event sensed in the atrium falls into the refractory period, the PMT is interrupted. The prolonged PVARP is up to 500 ms long.

Acute PMT may be dealt with by reprogramming (e.g., into the DDI mode) or by applying a magnet to turn off atrial sensing. If PMT episodes occur despite algorithms being enabled to inhibit the PMT, it is recommended that the PVARP be adjusted. If inefficient, the position and

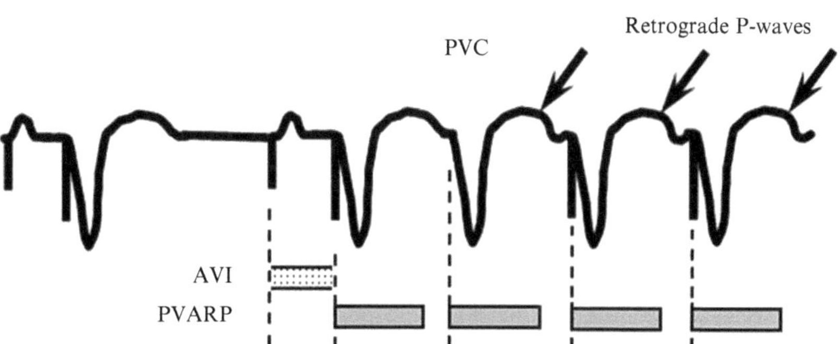

Fig. 9.14 Pacemaker-mediated tachycardia

operation of atrial and ventricular leads must be evaluated or pharmacological therapy must be considered to decrease the retrograde conduction [19, 20, 32].

9.9 Reducing Excessive Right Ventricular Pacing

Excessive right ventricular pacing may increase the risk of atrial fibrillation, left ventricular dysfunction, and congestive heart failure, in particular in patients with intact or irregular AV conduction [19]. One of the possibilities for reducing excessive right ventricular pacing is programming longer AV delays. In patients with intermittent or physical activity–dependent AV node block, intrinsic AV conduction may improve hemodynamic efficiency and increase the device's longevity by decreasing the number of ventricular pacing pulses.

The AV delay search function is devised to give preference to intrinsic AV conduction. It monitors intrinsic AV conduction that occurs after the end of a programmed AV delay. If the function is on, the AV delay is periodically prolonged (according to a programmed value) for a certain number of consecutive paced or sensed cardiac cycles. The search for AV delay remains active if the intrinsic PR intervals are shorter than the maximum programmed value of AV delay. The pacemaker returns to the programmed value of AV delay after the first ventricular pacing pulse or after the lapse of the search interval of certain number of cardiac cycles without sensing intrinsic ventricular activity.

The adaptation of pacing modes to the needs of patients with normal AV conduction is a more up-to-date solution preventing excessive right ventricular pacing. Instead of using a normal DDD(R) bradycardia pacing mode, in the case of AVIs with normal conduction, the device works in the AAI(R) mode at the LRL or a sensor-indicated rate [70]. Back-up VVI pacing is slower than the LRL by a certain value. When a loss of conduction is detected, the mode automatically switches to the DDD(R) mode for the restoration of AV conduction synchronization. The device evaluates slow ventricular contractions in a certain window as loss of conduction between atrium and ventricle. Slow contractions are defined as ventricular pacing or sensed ventricular activity that is slower than the AAI(R) pacing interval by a certain period. After switching into the DDD(R) mode, the device paces in the DDD(R) mode according to programmed parameters. When normal tracking ratios are restored, the pacing mode automatically switches back to the AAI(R) mode, with back-up VVI pacing. If reliable AV conduction is restored, ventricular pacing is not delivered as the back-up VVI mode runs in the background at a lower LRL. This function works together with ATR and adjusts the pacing mode according to the patient's atrial rhythm and the condition of AV conduction.

9.10 Types of Timing

A single-chamber pacing mode triggers an appropriate EI upon the detection of a sensed or paced event. If the pacemaker does not sense cardiac activity until the end of the EI, it paces the respective chamber. For example, if a ventricle is paced, the ventricular EI is timed from one ventricular event to another [69].

Dual-chamber pacemakers can have atrial, ventricular, or hybrid (combination of atrial and ventricular) timing [40]. The LRL is the rate at which the pacemaker paces the atrium, the ventricle, or both without sensed intrinsic activity or without sensor-driven pacing. During ventricular pacing, the EI is timed from one ventricular event to the following event. Any time an event is sensed in the ventricle, the timing changes from ventricular to atrial, or vice versa, according to the type of device. Then, the timing is ensured even with intrinsic AV conduction.

Regarding ventricular timing, which was assumed in the description of timing intervals, the AEI is fixed and is triggered by a sensed ventricular event. If a sensed ventricular event is detected during this interval, or rather during the AVI, it terminates the AEI and retriggers it again. If atrial pacing is followed by normal conduction of an impulse to the ventricles, and the interval between the atrial pacing and a sensed ventricular event (R wave) is shorter than the programmed AVI, the resulting pacing accelerates; that is, AV delay during ventricular pacing is equal to the programmed value, and AA and VV intervals are equal to the LRI. With normal AV conduction and a sensed ventricular event, the AV delay is shorter than the programmed value, and AA and VV intervals are shorter than the LRI. It also follows from the relationship between these intervals: AEI = LRI − AVI. Hence, if the AEI is triggered during the AVI, the resulting LRI is shorter.

As far as atrial timing is concerned, the atrial (atrio-atrial [AA]) interval (i.e., the interval between two atrial stimuli) is fixed. This AA interval is triggered by a sensed or paced atrial event. If activity is sensed in the ventricle during the AVI, ventricular pacing is inhibited, but the basic AA interval does not change. In other words, a ventricular event does not trigger the LRI. This is why atrial pacing is equal to the LRL upon a sensed or paced ventricular event. Hence, the resulting rate of this single-chamber-like atrial pacing does not change. For a premature ventricular event during the AEI, the AA interval, not the AEI, is set again. The system deducts the AA interval, adds the AVI, and thus attempts to cover a complete compensatory pause after ventricular extrasystole [40].

In dual-chamber pacemakers, an atrial event (sensed or paced) triggers the AVI that sustains the synchronization between atrial and ventricular events. If the pacemaker does not detect any sensed ventricular event until the end of the AVI, it paces in the ventricle. During synchronous AV activity, atrial timing is used in one of the manufacturer's systems. In the case of a loss of AV synchronization, that is, any time an event is sensed in the ventricle (e.g., spontaneous AV conduction occurs before the lapse of the AVI upon premature ventricular activity, Wenckebach activity, or mode switching), the system switches to ventricular timing. In another manufacturer's system, the EI is timed from one ventricular event to another during ventricular pacing. Any time an event is sensed in the ventricle, the ventricular timing changes to atrial. Precise pacing even during intrinsic AV conduction is ensured by this change in timing.

10 Implantable Cardioverter-Defibrillators

An implantable cardioverter defibrillator (ICD) is an active implantable medical device primarily used to deliver a high-energy electric shock to the heart via intracardial leads. The electric shock causes concurrent depolarization of all cell membranes in myocardial volume followed by an absolute refractory period. As a result, a tachycardia impulse spreading in the myocardium is interrupted. This implantable device also works as a pacemaker and applies antitachycardia pacing (ATP) to terminate ventricular tachycardia (VT) when high-energy shocks are not required.

Like in pacemakers, a literal code is used in ICDs to identify basic device modes (Table 10.1); the code contains only four characters. VVEV is the most common mode for single-chamber ICDs; VVED is used for dual-chamber ICDs.

High-voltage shocks in ICDs may be set up using a programmer and turned off by either using a programmer or applying a magnet above the ICD. Tachycardia functions may either be turned off completely, or detection may be turned on without a therapy, or both the detection and therapy may be turned on (the common long-term operation). The tachycardia function is switched off for safety reasons while handling the device and during implantation, end-of-life device replacement, or electrocauterization because handling the device, attaching leads using set-screws, etc., may cause artifacts that possibly could be interpreted as tachycardia. Following surgical interventions, in particular when the use of electrocautery is required, the ICD shock function must be switched off, and monitoring and an external defibrillator must be provided. In addition, electrocauterization may inhibit pacing in all types of implantable devices. If not equipped with the electrocauterizing (asynchronous) mode function, the device must be switched over to asynchronous pacing using a programmer or a magnet.

Requirements for ICDs are defined in *ISO 14708–6:2010: Implants for surgery – Active implantable medical devices – Part 6: Particular requirements for active implantable medical devices intended to treat tachycardia (including implantable defibrillators)* [72]. The standard provides technical information on measuring parameters, susceptibility to electromagnetic interference, etc.

10.1 The Design of Implantable Cardioverter-Defibrillator Systems

An ICD is similar in design to a pacemaker. Mechanically, it comprises a titanium can and a polymer header. In certain types, even an antenna used for radio communication with a programmer or a remote monitoring unit may be visible. The metal can contains, in addition to a battery and electronic circuits, comparatively large high-voltage electrolytic capacitors with the capacity range of 60–140 μF (according to the manufacturer) necessary for energy accumulation before delivery of a defibrillation shock. Maximum accumulated energy is currently around 40 J. Capacitors, together with the battery, make up most of the ICD's volume (Fig. 10.1). In terms of electronics, the ICD contains only a monolithic integrated circuit and several external components [73].

If a longer period of time passes between ICD charging cycles, dielectric material re-formation in high-voltage capacitors may dissipate. This may prolong the charging period. For that reason, high-voltage capacitors should be charged on a regular basis, either automatically or at regular follow-ups by re-formation using a programmer. High-voltage capacitors thus are charged to maximum voltage; after charging, the voltage gradually decreases, and, depending on the value, dissipates within several minutes.

D. Korpas, *Implantable Cardiac Devices Technology*, DOI 10.1007/978-1-4614-6907-0_10,

Table 10.1 NASPE/BPEG defibrillator codes [71]

Position	I	II	III	IV
Category	Shock chamber	Antitachycardia pacing chamber	Tachycardia detection	Antibradycardia pacing chamber
Letters	0 – None A – Atrium V – Ventricle D – Dual (A+V)	0 – None A – Atrium V – Ventricle D – Dual (A+V)	E – Electrogram H – Hemodynamic	0 – None A – Atrium V – Ventricle D – Dual (A+V)

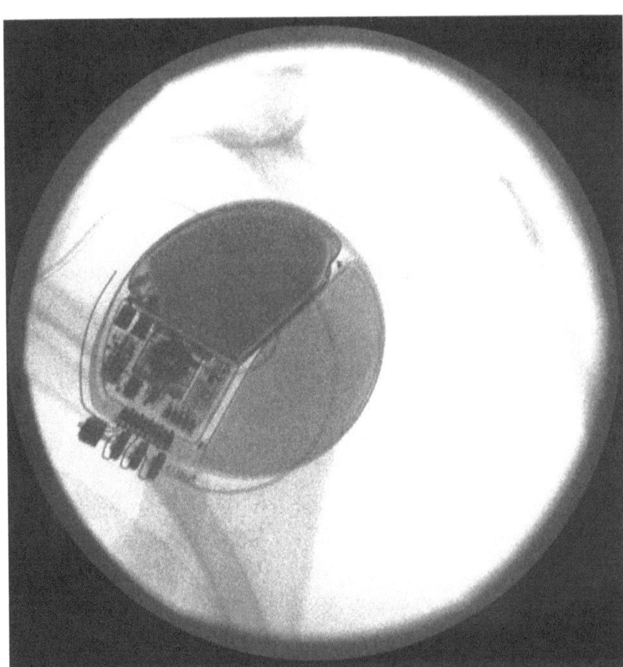

Fig. 10.1 X-ray image of an implantable cardioverter-defibrillator

Table 10.2 Example of ventricular detection zone configuration [82]

Zone configuration	VT-1 zone	VT zone	VF zone
1 zone	–	–	200 beats/min
2 zones	–	160 beats/min	200 beats/min
3 zones	140 beats/min	160 beats/min	200 beats/min

10.2 Tachycardia Detection

The application of appropriate treatment depends on a precise classification of the patient's rhythm. The ICD system evaluates the heart rhythm based on individual cardiac cycles. After a sensed event, the cycle length is measured and compared with programmed detection parameters. Refractory periods, together with noise windows, may eliminate sensing of signals other than physiological ones and prevent the delivery of potentially undesirable therapy. Atrial refractory periods in DDD(R) and DDI(R) modes, right ventricular refractory periods after a right-ventricle sensed event or after charging capacitors, and refractory periods after shock may be nonprogrammable.

A VT therapy zone is a range of heart rates in a frequency domain delimited by at least one programmed VT threshold. A programmed VT threshold value is a value to which the ICD compares each sensed cardiac cycle period (Table 10.2).

As a rule, as many as three VT zones may be programmed; one of them may be programmed as a monitoring zone (without a therapy, only for tachycardia diagnostic recording). For each zone, a separate therapy prescription can be set up. They are designated according to expected tachycardia (e.g., VT-1, VT, and VF). The zone setting is interrelated with basic bradycardia parameters. The lowest VT zone threshold value must be higher by a determined difference than the maximum tracking rate, maximum sensor rate, and lower rate limit parameters.

The ICD device must be capable of distinguishing between several types of ventricular arrhythmias. Ventricular fibrillation (VF) is a very fast rhythm with a low amplitude and irregular cardiac intervals. VT is slower compared with VF, with regular intervals. Supraventricular tachycardia is a fast rhythm originating in atria; the rhythm is not indicated for a ventricular therapy. Theoretically, atrial arrhythmia can also be achieved by a shock applied between the lead proximal defibrillation electrode and the device can. Nevertheless, this method is rarely applied in practice because it is painful to a fully conscious patient. After delivery of therapy, the ICD must evaluate the patient's rhythm again, and, in the case of persisting arrhythmia, apply another therapy. Even after the termination of the event, the device must continue monitoring possible recurrence of tachycardia.

Steps to detect VT are as follows:

- Initial detection
- Reconfirmation
- Redetection and detection after shock

10.2.1 Initial Detection

Criteria for initial detection of VT consist mainly of programmable parameters for the detection zone threshold rate and duration. The detection criteria may also be extended by some enhancing functions. These may be applied at the initial and postshock detection of VT for the purpose of distinguishing VT from supraventricular tachycardia. The functions extend the detection specificity beyond the limits of the basic parameter (i.e., the zone threshold rate) because applying shock treatment in case of supraventricular tachycardia it is not necessary or even apposite.

The ICD triggers a VT therapy if detection conditions are met. It happens upon the concurrence of all of the following situations:

- The condition of a zone detection window is met and remains met throughout the period determined by the duration parameter.
- The zone duration time determined by the duration parameter elapses.
- The condition of a higher zone detection window is not met.
- Detection-enhancing functions indicate a therapy.
- The last sensed interval is in the tachycardia zone.

If the above-listed criteria are not met, the therapy is not delivered, and the ICD continues evaluating intervals.

A detection time window is used to ensure appropriate treatment application. Each tachycardia zone has a detection window comprising a certain number of RR intervals last measured by the ICD. Upon measurement, each new interval is compared with a programmed zone threshold rate and classified as fast or slow (i.e., above or below the zone threshold rate, expressed in milliseconds) within each detection window. The ICD gets prepared for a potential episode if it counts a defined number of consecutive fast intervals. The condition of a detection window remains met if certain portion of the intervals (e.g., six of ten) remains classified as fast. If the number of fast intervals drops below this limit, the condition of a zone detection window ceases to be met – it would be considered to be met only if the defined number of intervals was classified as fast again. Since a higher zone threshold must be programmed to a value higher than a lower zone threshold, the interval classified as fast in a higher zone will also be classified as fast in all lower zones [70, 74]. Some defibrillation systems make use of a timer duration parameter, measuring the period of time for which the rhythm in individual zones must be constant before a therapy is applied. The duration timer is started if the condition of a respective zone detection window is met. Programmed duration time and its passing are evaluated after each cardiac cycle. As the timer counts synchronously with the cardiac cycle, the programmed duration parameter may be exceeded by one complete cardiac cycle. If the condition of a zone detection window remains met, the timer continues counting. If the last detected interval after the lapse of the duration time is still in the respective detection zone, detection conditions are considered to be met, the application is not suppressed by another programmed detection enhancing function, and a therapy is started. On the condition that the last detected interval is not in the respective zone, a therapy is not delivered. Each following interval will be checked until an interval falling into the original zone is identified or until the window condition is met. A separate duration time parameter is programmed for each tachycardia detection zone. The value of the duration time parameter programmed in lower detection zones must be equal to or higher than the value in higher ventricular zones. Longer durations may be used to prevent delivering a therapy at inconstant tachycardia.

Duration timers count independently of one another in their respective detection zones. If arrhythmia is detected in the highest zone, its timer gains priority over timers in lower zone. If the duration parameter time elapses and the detection condition is met, a therapy corresponding to the respective zone is applied regardless of whether the timer in the lower zone has finished counting. If the condition of a higher zone detection window ceases to be met, lower detection zone timers are taken into account again.

10.2.2 Redetection

After application of therapy, the ICD continues evaluating the heart rhythm to identify the need for another therapy. Redetection criteria are applied as for the initial detection. If the redetection criteria are met, the therapy to be applied is selected. Parameters for duration of ventricular redetection duration and duration aftershock are applied to identify tachycardia during ventricular redetection. To minimize the time until possible application of the therapy, it is recommended that the redetection duration in VT-1 and VT zones in multiple-zone configurations be programmed to ≤ 5 s.

Redetection can be started by both the previous therapy application and a therapy suspended because of reconfirmation, a therapy suspended manually from a programmer, or detection conditions being met without a programmed therapy. Redetection makes use of the same detection window and programmed detection zone process that is used for tachycardia identification during the initial detection. Differences between the initial detection and redetection lie in varying duration parameters and available detection enhancing functions.

Upon delivery of ATP, the ICD monitors the heart rhythm after each burst and makes use of the ventricular redetection window and redetection duration parameter to identify the end of arrhythmia. The ATP will continue delivering bursts in programmed order until the redetection recognizes that the therapy has been successful, a set number of ATP bursts in the sequence has been applied, or a transition into another zone has occurred. Canceling an ATP burst terminates the entire running ATP sequence. Upon application of shock therapy, the ICD monitors the heart rhythm after each shock and makes use of ventricular detection windows and post-shock detection enhancing functions to identify the end of arrhythmia. While charging the ICD capacitor, the ICD continues sensing arrhythmia. A shock therapy continues until redetection recognizes that the therapy has been successful (the episode ends), all ventricular shocks available in the respective zone have been applied, or a transition to a lower zone occurred.

10.2.3 Detection-Enhancing Functions

For the purpose of restricting false-positive shocks, when a shock therapy would be applied – even for arrhythmias not originating in heart ventricles –ICD systems usually include algorithms extending the initial detection. These enhancing functions evaluate detected tachycardia as ventricular or, for example, tracked atrial arrhythmia. Sensitivity to VT of 100 % and maximum specificity is required from the functions. The absolute sensitivity requirement means that all VT episodes are detected, despite their cause. The specificity means that only real VT and not tracked supraventricular tachycardia are treated.

There are several functions, and each manufacturer implements their own approach. In general, algorithms may be classified according to how they evaluate arrhythmias. The evaluation may be based on a more detailed analysis of sensed tachycardia time intervals, comparison of atrial and ventricular rhythm, or the evaluation of electrogram morphology. These functions are usually not available in the highest detection zone.

10.2.3.1 Algorithms Based on a Time Interval Analysis

A classic approach available in the majority of ICDs is the use of onset and stability parameters analyzing cardiac cycle intervals [70, 74, 75]. The onset parameter identifies onset of tachycardia as sudden or gradual; stability quantifies the course of tachycardia, that is, the variability of the RR interval.

The onset function distinguishes between sinus tachycardia starting slowly and pathologic ventricular tachycardia starting suddenly. It measures the rate of change from a slow ventricular rhythm to tachycardia. The onset function may be programmed as a percentage of cycle length or as the interval length. The onset value represents a minimum difference between intervals (or a group of intervals). If the onset parameter is programmed to a value lower as a percentage, the ICD will need greater acceleration of cardiac action to identify the rhythm as VT. If the rate increases gradually, then the onset function avoids sensed intervals from being interpreted as VT. To reduce the number of shock therapies, the onset function must be programmed to a higher numerical value.

Stability analysis function distinguishes between irregular and regular ventricular rhythms. The evaluation is based on measuring the variability of the RR interval during tachycardia. Certain tracked supraventricular tachycardia may lead to ventricular acceleration up to the detection zone. These tachycardia relating to irregular heart rate have higher variability of the RR interval when compared with real monomorphic VT, which is stable. The function may also be used to distinguish monomorphic and polymorphic VT. The stability algorithm calculates differences in the length of the RR interval. These differences are determined throughout the duration period and, finally, an average difference is calculated. Then, rhythm stability is evaluated by comparing the average difference with a programmed stability threshold. Another system classifies an interval as instable if the difference between the interval and any other of three previous intervals is bigger than the programmed interval for stability. If the difference exceeds the programmed threshold value, the ventricular action is not stable and the rhythm is reported as instable (supraventricular tachycardia). To reduce the number of shock therapies, the stability function must be programmed to a lower numerical value.

Onset and stability functions may be combined in some manufacturers' ICDs to ensure higher specificity of tachycardia classifications. The function combination may be programmed so that a therapy is indicated at concurrent validity of onset and stability functions or the validity of only one of them.

10.2.3.2 Morphological Algorithms

These functions work on the basis of mathematical analysis of intracardial signals sensed at tachycardia or comparison of the sensed rhythm characteristics with a stored supraventricular tachycardia template. The description of two such detection-enhancing functions may serve as an example; according to data from clinical studies, the functions show high specificity [76] while maintaining maximum sensitivity.

The algorithm of one manufacturer makes use of a sensed electrogram wavelet transformation for the purpose of dynamically comparing tachycardia morphology with a previously stored sinus rhythm template, which is updated continuously. The algorithm stores the template of sensed QRS complexes of the intrinsic rhythm and creates a template comprising 48 electrogram samples taken every 4 ms using HaaR wavelet. The template morphology transformation is compared in real time with transformation of electrogram morphology during tachycardia. This comparison is expressed as a percentage agreement describing the level of morphological similarity of tachycardia and sinus rhythm electrogram. Sensed QRS complexes with agreement under a programmed threshold (nominally 70 %) are identified as VT. If at least three of the past eight QRS complexes comply with the stored template, the algorithm suspends the therapy [77].

Another system makes use of analysis of timing and difference of the resulting electrogram vector for various rhythms. The ICD first analyzes whether the ventricular rate is higher than the atrial rate (only in dual-chamber ICDs and cardiac resynchronization therapy defibrillators). If so, a therapy is initiated. If the ventricular rate is not higher, an evaluation is started. At the normal sinus rhythm used as a template, electrical activity passes from the atrium to the ventricles. Even supraventricular tachycardia has a similar electrogram. In contrast, VT originates in the ventricular myocardium, and thus the resulting vector of electrical activity differs. The system senses electrograms both by lead pacing electrodes for the near-field and lead shock electrodes in the far-field. The electrogram from the pacing electrodes is used for time comparison with the far-field electrogram and a correlation analysis of stored and currently sensed rhythms. Amplitudes of eight prescribed instants of time are compared. The correlation threshold (nominally 94 %) may also be adjusted. If a correlation value is lower than or equal to the threshold value, the cardiac cycle is regarded as VT. The template of a patient's intrinsic sinus rhythm may be recorded both in a doctor's office via a programmer and automatically at appropriate instants [70].

10.3 Tachycardia Therapies

An ICD may deliver two different types of therapies for the purpose of terminating VT or VF: either ATP or defibrillation electric shocks. ATP therapies are bursts applied between defibrillation lead pacing electrodes. Shocks are high-voltage biphasic or monophasic pulses delivered via defibrillation shock electrodes.

The type of therapy to be applied in a ventricular detection zone is subject to ventricular therapy prescription. A zone prescription, usually with the exception of the highest VF, may comprise only ATPs, only shocks, or their combination. Each ventricular zone may be programmed according to a separate ventricular therapy prescription. The therapy intensity must, however, always have a nondescending tendency within each zone. In the lowest zone of the multiple-zone configuration, some or all therapies may be turned off, and tachycardia in this zone are only recorded in the ICD's memory. ATP therapies are regarded as therapies with the same intensity; it is, however, lower than the intensity of any shock. Shock energy is determined by its programmed value. In the multiple-zone configuration, therapies may have lower, higher, or equal intensity in a higher ventricular zone, depending on the type of device, with respect to therapies in a lower ventricular zone. In most contemporary ICDs, a process referred to as ATP is implemented before or during capacitor charging; even in the highest VF detection zone, the application of an ATP burst is tried before a shock is delivered in attempt to terminate tachycardia painlessly.

Based on initial ventricular detection criteria, an ICD chooses the first prescribed therapy in the zone in which tachycardia is detected. Upon its delivery, redetection of arrhythmia conversion commences. If the arrhythmia is not converted but is redetected in the same ventricular zone, the following programmed therapy in the respective zone is selected and delivered, and redetection is carried out. If arrhythmia is redetected after the application of therapy in a higher or lower detection zone, a therapy with an intensity equal to or higher than the intensity of the previously applied therapy is selected and applied from the new zone prescription.

Upon application of a shock, heart contractions must be sometimes restored by the right ventricular pacing. The pacing delay interval after a shock specifies the possible beginning of pacing upon application of a ventricular shock. The timing of the initial pacing pulse in the interval depends on the intrinsic cardiac activity during the delay. Provided that R waves are sensed during such a delay, the device applies pacing only if the action sensed is slower than the lower rate limit after therapy. Further pacing pulses are applied as needed according to the pacing prescription.

10.3.1 Antitachycardia Pacing

ATP represents the possibility of a VT episode therapy and restoration of the patient's normal sinus rhythm primarily in the case of monomorphic VT (Fig. 10.2). It is assumed, based on clinical studies, that ATP is efficient for VT up to the rate of 250 beats/min. ATP consumes less battery energy than a shock and is better tolerated by patients. During ATP, a sequence of precisely timed pacing pulses is applied.

Burst, ramp, scan, and their combinations (ramp + burst/scan) are basic ATP schemes. Burst may refer either to a basic ATP sequence or one cycle of each sequence including a certain number of pulses. ATP prescription can be programmed individually by setting the number of bursts to be applied, the number of pulses in each burst, a coupling interval, burst length, and a minimum pacing interval. Amplitude and width of the ATP pulse are common for all schemes and are programmable independent of normal pacing settings.

The coupling interval controls the timing of the first burst pulse. It defines the time between the last sensed event that meets the detection criteria and delivery of the first burst pulse. If an adaptive coupling interval is programmed, the parameter is adjusted according to the patient's rhythm based on the average value calculated from the last several cardiac cycles.

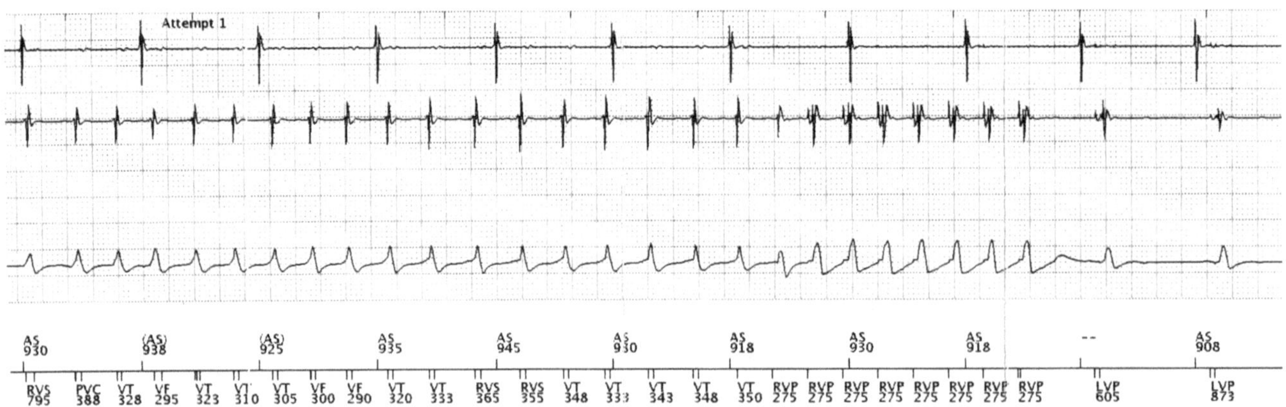

Fig. 10.2 Successful antitachycardia pacing (burst)

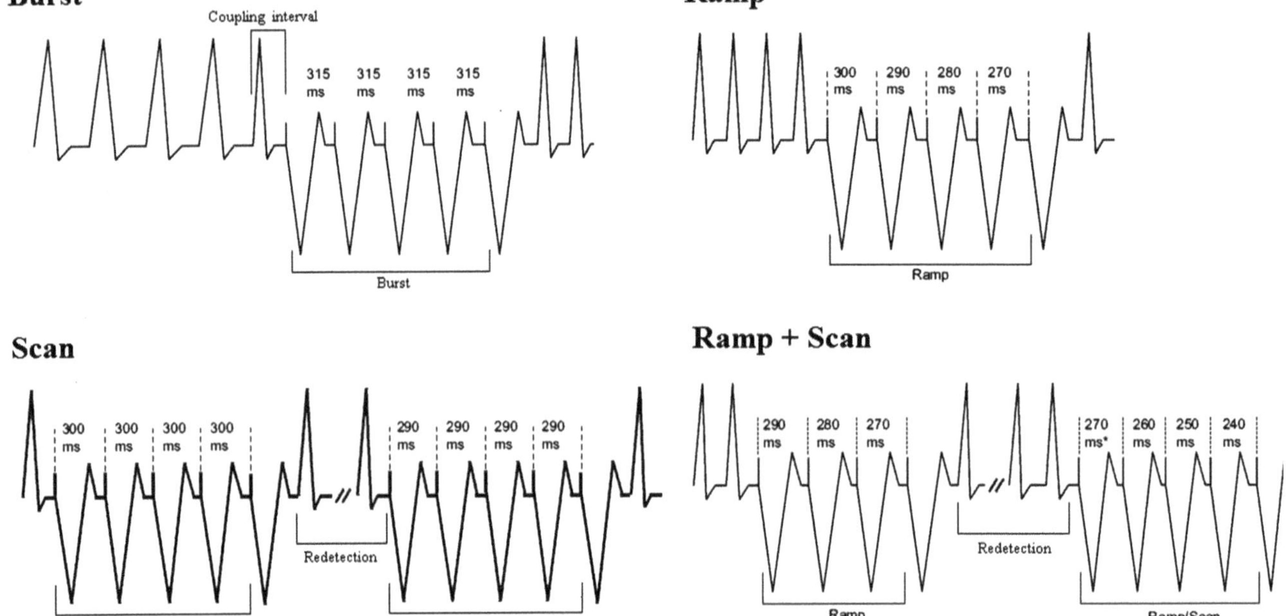

Fig. 10.3 Antitachycardia pacing therapy schemes [82] (© 2012 Boston Scientific Corporation or its affiliates. All rights reserved. Used with permission of Boston Scientific Corporation)

An ATP is a sequence of precisely timed pacing pulses aiming to interrupt the reentry loop; the pulses are applied at a rate higher than a patient's tachycardia heart rate (Fig. 10.3). It is characteristic of the burst ATP that all pacing intervals in a burst are constant. In contrast, in the case of the ramp ATP, the interval of consecutive pacing pulses in a burst gradually decreases over time. When applying individual pacing pulses following a burst, the respective interval is shortened by a programmed shortening parameter value until the last pacing pulse in a burst or the minimum interval value is reached. The scan ATP is a sequence of pacing pulses, where the ATP burst coupling interval decreases systematically. Ramp and scan ATPs can be combined in certain ICDs.

10.3.2 Shock

An ICD applies shocks synchronously or asynchronously in response to a sensed ventricular event. A shock has several programmable and nonprogrammable parameters. The programmable parameters include energy, polarity, a shock vector, and, in certain ICDs, specific defibrillation waveform parameters, as mentioned below.

Shock energy is programmable to values determined by program application, from 0.1 J to the device maximum. Time required for charging capacitors to a programmed level depends on the programmed energy, battery status, and the time elapsed since the previous shock or the last capacitor

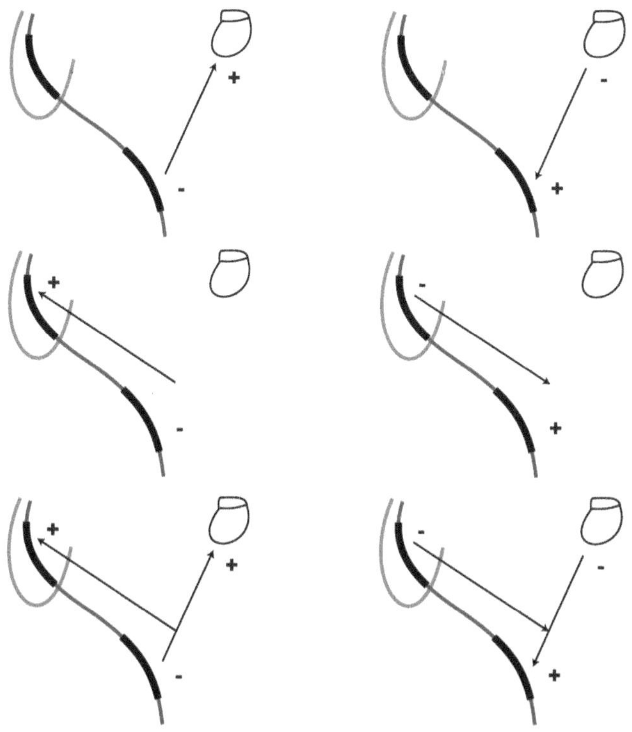

Fig. 10.4 Defibrillation configurations [82] (© 2012 Boston Scientific Corporation or its affiliates. All rights reserved. Used with permission of Boston Scientific Corporation)

re-formation. During idle time, a capacitor may be deformed, which may result in the charging time being prolonged slightly. The shock energy remains constant throughout the longevity of the device, regardless of changes to the lead impedance or battery voltage.

The waveform polarity refers to the relationship between values of the voltage of the leading edge of the defibrillation pulse at the output of the lead defibrillation electrode. Changing the polarity by physically swapping a lead cathode and anode on the ICD output is prohibited. A programmable function must always be applied. Switching the physical polarity may result in damaging to the device or postoperative nonconversion of the arrhythmia.

When using a defibrillation lead with two coils (right ventricular coil [RVC]; supraventricular coil [SVC]), three shock vector configurations are programmable (Fig. 10.4):

- RVC–can. This vector makes use of the ICD metal can as an active lead. The energy passes from a distal shock electrode to the ICD can via only one pathway. This configuration must be chosen when using a lead with one defibrillation electrode (single coil).
- RVC–SVC. The vector does not make use of the ICD can as an active lead. The energy passes from a distal shock electrode to the proximal electrode. This vector must never be used with a single-coil lead because a shock would fail to be delivered.

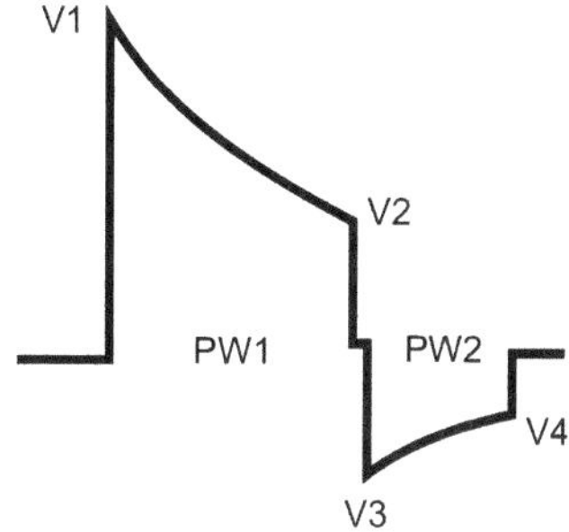

Fig. 10.5 Defibrillation waveform

- RVC–SVC–can. This vector makes use of the ICD metal can as an active electrode in combination with a dual-coil defibrillation lead. Current density creates a double pathway from the distal electrode on the shock lead to the proximal electrode and the ICD can.

Defibrillation waveform parameters are crucial for the efficacy of defibrillation. In principle, we distinguish between a monophasic and biphasic defibrillation waveform with characteristic values, per Fig. 10.5. Several characteristic values may be identified in the defibrillation waveform. In the amplitudes:

- Positive phase leading-edge voltage V1 (V)
- Positive phase tilt V2 (%)
- Negative phase leading-edge voltage V3 (V)
- Negative phase tilt V4 (%)

In the time domain:

- Positive phase duration PW1 (ms)
- Negative phase duration PW2 (ms)

Relative values of these parameters vary with individual manufacturers. Absolute values are determined by set shock energy, defibrillation circuit impedance, and nominal capacity of the ICD capacitor. Positive phase leading-edge voltage V1 at maximum energy exceeds 700 V. The greater the capacity of the capacitor, the longer the PW1 time (the more time is needed for discharge).

Manufacturers apply two methods to express shock energy: delivered or stored. The delivered energy is the energy that passes through output shock connectors and dissipates in a patient or par value resistance (50 Ω). Delivered energy will always be lower than stored energy because a portion of the stored energy dissipates even in the ICD. In addition, shocks can be terminated before the entire energy stored in the capacitors is transferred. The delivered energy is a crucial quantity for the determination of the ICD's efficacy.

10.4 ICD Diagnostic Features

The ICD automatically records information on the detection and therapy for each detected tachycardia episode. Then, the data can be searched through at various levels of detail using a programmer. For each episode, the data stored includes information on episode time, electrograms with annotated markers, cardiac intervals, etc. The information comprises an episode number, date, time, a type (VF, VT, spontaneous/commanded), average atrial and ventricular rates, the type of therapy delivered and its parameters, and episode duration. The device stores the maximum number of episodes of individual types until the memory is full. As soon as the memory capacity available for storing episode data is filled, the device rewrites older stored episodes according to their priority.

The trend function provides graphical representation of bradycardia counters and histograms. This data can be useful for the evaluation of a patient's condition and efficacy of programmed parameters. Atrial and ventricular markers, the patient's daily activity rate, time of the ATR mode switching, PMT, daily breathing rhythms, intrinsic signal amplitudes, and pacing and defibrillation circuit impedances are displayed, depending on the type of device.

Histograms contain the total number and percentage of paced and sensed events for the respective heart chamber. Histogram data can provide clinical information, such as frequency distribution of the patient's intrinsic heart rhythm and the proportion of paced and sensed events according to heart rate.

A patient may commence recording electrograms, intervals, and annotated markers during a symptomatic episode by placing a magnet on the device. If the function is activated, the device stores a record of arrhythmia from a certain interval before the recording is started. This is made possible because electrograms are moved in the register so they are still available in the memory after the magnet is applied. The data stored include an episode number, atrial and ventricular action at the time of magnet application, and the time and date of the beginning of magnet application. However, upon the activation of this function, all other functions of the magnet, and primarily the inhibition of therapy, are deactivated. Before the magnet is given to a patient, it must be evaluated carefully to determine whether the function of recording is really activated. If the magnet function remained set to therapy inhibition unintentionally, it could have serious consequences later.

10.5 Electrophysiologic Testing Using an ICD

Electrophysiologic testing functions allow noninvasive induction and termination of arrhythmia. They are commonly used during ICD implantation to test the efficacy of set detection and therapy.

Common methods of VT induction include the following:

- A T wave shock
- Programmed electrical stimulation (PES)
- A burst

Some systems also allow large-amplitude pacing pulses (9 or 15 V) applied via lead shock electrodes. The functions (PES, a burst) can also be available in an atrium with backup ventricular pacing.

During electrophysiologic testing, the status of ICD detection and therapy processes is displayed in real time on the programmer. If an episode is in progress, the zone in which the respective detection condition is met is indicated. During induction, the device detection is switched off automatically; after termination of induction, the detection is switched on again automatically. When applying any type of induction, the ICD stops performing other activities until the end of the induction, after which the detection is applied and the ICD responds in an appropriate way. A patient may be medicated before the induction. During electrophysiologic testing, patients must be carefully monitored, and an external defibrillator must be available because induced VT may degenerate to VF. Once induction is commenced by any method, the application does not stop, even in the case of dropped telemetric communication.

When inducting tachycardia by means of a T wave shock (Fig. 10.6), relevant programmable parameters must be set.

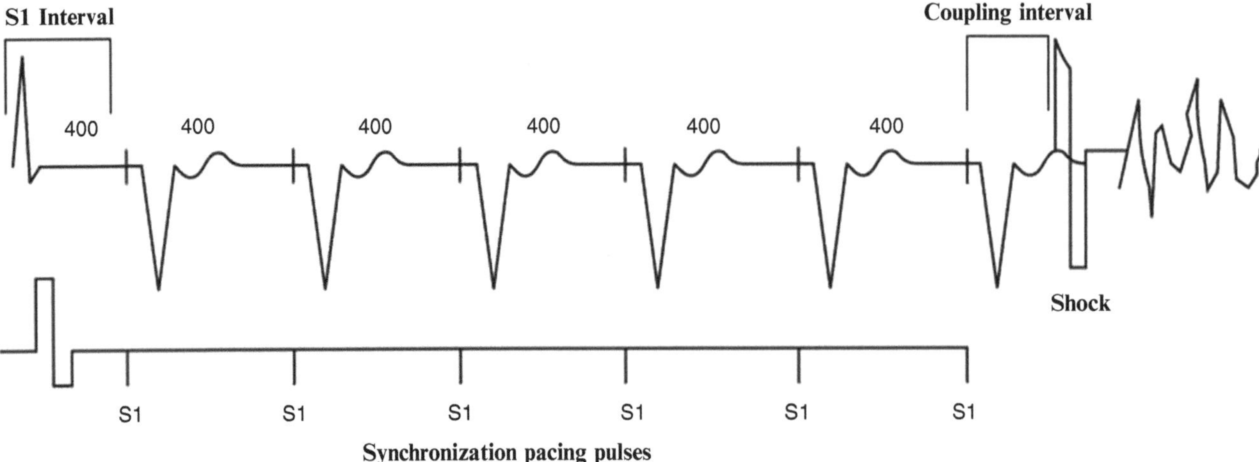

Fig. 10.6 Shock on T-induction method

Before delivery of the T wave shock, synchronization is carried out by an asynchronous burst V00 so that T wave timing could be anticipated. According to a set coupling interval or delay after the last synchronization pulse, a shock is delivered, the timing of which is assumed in the cardiac cycle vulnerable phase. T wave shock energy is set in the range 1–2 J.

PES allows the ICD to deliver an optional number of pacing pulses in regular time intervals (S1) followed by premature stimuli (S2–Sx) inducing or terminating arrhythmias [74, 75]. Excitation pulses (or pulse S1) serve for management of pacing and cardiac rhythm at a rate higher than the intrinsic action. Thus, precise synchronization of the stimuli with the cardiac cycle is ensured. The initial pulse S1 is coupled with the last sensed or paced event by interval S1. Pulses are applied gradually in asynchronous modes A00 or V00 according to a programmed prescription until a pulse is encountered that is set to off. The induction of PES is complete after an excitation drive train and extra stimuli are applied and after ICD detection is restored automatically. PES can be applied to induce both atrial tachycardia and VT.

Burst induction is applied to induce or terminate arrhythmias; it allows various types of pacing, possibly applied asynchronously, either in an atrium or a ventricle. A burst with optional cycle length, gradual acceleration, or set at 50 Hz, can be selected depending on the manufacturer and the type of ICD.

Commanded (manual) therapies are started from the programmer. Available types include a defibrillation (cardioversion) shock and all types of ATP. Commanded therapies may be applied independent of programmed detection and therapy parameters. If a device is applying a therapy and a commanded method is initiated, the electrophysiological function has a higher priority and suspends the therapy or detection in progress. The commanded shock function enables the delivery of a shock with adjustable energy and coupling interval. The shock waveform and its polarity are identical to shocks after the initial detection. Upon the delivery of the commanded shock, redetection is applied, and postshock pacing is activated, if programmed. Commanded ATP also allows the application of schemes independent of programmed detection and therapy parameters [70, 77]. During electrophysiologic testing, temporary pacing settings can be used to secure a patient. This may also be applied during a patient's routine follow-up for the evaluation of the efficacy of delivered treatment.

10.5.1 Determining the Defibrillation Threshold

After lead connectors are connected to the ICD and the device is placed into an implantation pocket, electrograms as well as lead signals must be evaluated in real time using the

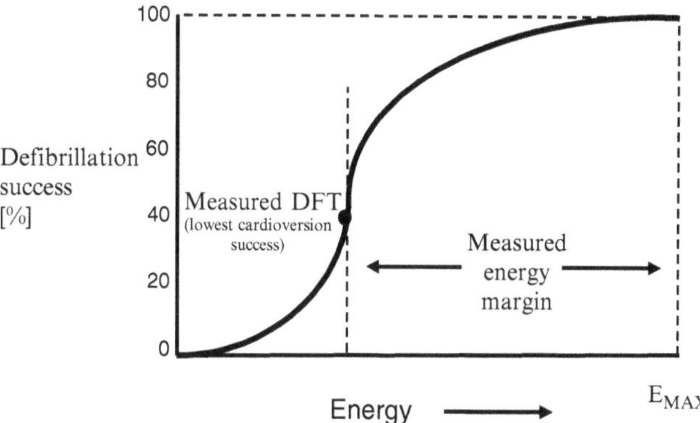

Fig. 10.7 Relationship between successful probability defibrillation and shock energy

implanted device. The signal from the implanted lead should be uninterrupted and without artifacts, similar to a surface electrocardiogram. An interrupted signal may be a sign of breakage or other damage to the lead or defective insulation. To obtain acceptable lead signals, the configuration and position of implanted leads must be evaluated for the suitability for each patient and the sufficiency of programmed shock energy or maximum energy shock for reliable tachycardia cardioversion.

Testing a defibrillation threshold (DFT) allows a safety reserve for a treatment shock to be determined. The safety reserve is calculated as the maximum energy of the ICD minus DFT, where DFT is the minimum tested energy that reliably turns VF or polymorphic VT into a sequence of test arrhythmias with descending defibrillation energy. DFT refers to the energy level at which arrhythmia is last converted (Fig. 10.7). Such testing is carried out by inducing arrhythmias using a programmer and by delivering shocks to the patient's heart from the ICD via defibrillation shock electrodes. During a shock or delivery of ATP, conducting objects or devices must be prevented from making contact with the system because the energy could be diverted, resulting in a decrease in the amount of energy delivered to the patient and subsequent possible damage to the implanted system. It is important that the cardioversion should be carried out as soon as possible after the arrhythmia is induced because longer-lasting tachycardia is more difficult to terminate.

Reliable conversion should be proved at a lower energy level than the device's maximum shock energy. Technically, it would be ideal to identify a real defibrillation threshold, that is, the minimum amount of energy required to terminate VF. Because repeated action represents a considerable load for a patient, each shock must be regarded as proarrhythmogenic, and, since the defibrillation threshold depends on many circumstances, energy 10 J lower than the device's maximum shock energy is often tested. Between induced arrhythmias, the patient's starting blood pressure and electrophysiological condition should be restored regardless of whether the cardioversion has been successful. In addition, it is recommended to set an interval of at least 1 min between induced arrhythmias; 5 min is, however, more common. If the version is unsuccessful, the patient must be resuscitated by means of an external defibrillator. After an unsuccessful shock, only limited possibilities to improve the successfulness of defibrillation shock are available. Shock polarity, shock waveform (monophasic or biphasic), and the configuration of lead defibrillation electrodes can be changed. This includes switching off a proximal shock electrode. Upon improper sensing of cardiac action, delayed detection, or detection failure due to signals with a low amplitude, the lead position must be changed or a separate lead must be used only for sensing. If reliable arrhythmia conversion cannot be reached by any means using an implanted endocardial defibrillation lead, additional implantation of a subcutaneous lead, SQ array, or another lead system – and carrying out another conversion test – is required.

The defibrillation capability, that is, the defibrillation threshold value, depends on many circumstances. Clinical studies were conducted suggesting that in situ electrical field intensity of 5.4 V/cm must be applied with a biphasic shock to reach 87 % success for defibrillation or 3.5 V/cm to reach 80 % success [78]. Other studies [79] recommend applying an electrical field intensity of 5 V/cm to more than 90 % of myocardial volume to achieve successful defibrillation. Upon delivery of a biphasic shock, with the defibrillation system shock electrode voltage of approximately 700 V (with a set maximum energy) the electrical field intensity is absolutely sufficient. The DFT is individual, though, and there are cases when the configuration of the defibrillation system must be adjusted.

In the past, during implantation of endocardial defibrillation leads, the defibrillation threshold was evaluated even before connecting to an ICD using an external cardioverter-defibrillator (ECD). ECDs were external defibrillation systems capable of delivering powerful shocks (up to approximately 40 J) into a patient's body through an implanted lead. With the ECD, it was necessary to pay attention to the proper connection of lead connectors to corresponding connectors on the ECD system. If a high-voltage defibrillation shock was applied through a pacing lead, the heart could be injured.

10.6 Concurrent Activity of an ICD and a Pacemaker

In some cases, the ICD must be implanted in a patient with a previously implanted pacemaker. Pacemakers may affect the function of an ICD and interfere with detection of tachycardia. If a pacemaker is not inhibited by arrhythmia during tachycardia and paces, the pacing pulse may be detected by the ICD and may interpret the pacemaker rate as the normal rhythm. As a consequence, the ICD would fail to identify the arrhythmia and would not deliver a therapy. A pacemaker may also be a source of signals sensed by the ICD and occurring as a result of noise, lead dislocation, or inefficient pacing. The ICD can then measure a faster than actual action rate and apply a therapy inappropriately. For these reasons, the concurrent use of a pacemaker and an ICD is not recommended. Furthermore, the use of unipolar pacemakers with ICDs is contraindicated.

If the coexistence of both systems is inevitable, ICD sensing leads must be as far as possible from pacemaker pacing leads. After lead implantation, signals from ICD sensing leads must be evaluated, and minimal interaction with artifacts from a pacemaker must be ensured. For the purpose of identification of pacemaker artifacts, a system with a minimum band width of 2,000 Hz is recommended. The artifacts can be eliminated by replacing the leads. The distance between sensing lead electrodes must not exceed 1–2 cm; otherwise the signal could exceed a device refractory period, which causes too high a level of sensing at the normal rhythm or too low a level of sensing at a polymorphic rhythm. In addition, a large distance between lead pacing/sensing electrodes may contribute to too high a level of sensing because a large repolarization signal (T wave) will be introduced.

Because the significance of pacemaker artifacts and possible electrograms of various tachycardia can hardly be anticipated, these artifacts should be minimized. The pacemaker should be programmed to the lowest amplitude permissible for safe pacing, maximum sensitivity for the inhibition of pacing at ventricular fibrillation, and minimum acceptable pacing rate. In addition, the ICD bradycardia pacing function may be switched off, or the function can be programmed to a lower rate than is set in the separate pacemaker. At the same time, switching off postshock pacing in the ICD should be considered if a pacemaker has sufficient output pulse parameters [70].

11 Cardiac Resynchronization Therapy

Cardiac resynchronization therapy (CRT) is a nonpharmacological method used to treat chronic heart failure. One symptom of heart failure is ventricular dyssynchrony decreasing the stroke volume. CRT enables treatment of ventricular dyssynchrony by means of delivering pacing pulses to the right and left ventricle. As a consequence, a better heart contraction mechanism and an increase in cardiac output is achieved. The clinical efficiency of CRT was confirmed in large randomized studies, which are described in the Chap. 6, dealing with indications for treatment. CRT improves hemodynamic parameters and quality of life of patients with severe heart failure. The first CRT systems made use of a left ventricular (LV) lead inserted via an epicardial approach requiring thoracotomy. Today, the lead is inserted via the subclavian vein through the coronary sinus and into the target coronary vein. The pacing electrodes of the LV lead should be suitably located at the place of the most recent ventricular activation during intrinsic cardiac impulse conduction. In most patients, this site is assumed to be on the left ventricle lateral wall if the lateral or posterolateral vein is available.

Biventricular pacing may be delivered using both pacemakers (CRT pacemakers [CRT-Ps]) and defibrillators (CRT defibrillators [CRT-Ds]). Technically, pacing components of CRT-Ds are identical to those of CRT-Ps [80, 81].

11.1 Securing Left Ventricular Pacing

Biventricular pacing was first applied in a human as early as the 1970s in the framework of attempts to suppress ventricular arrhythmia by overdriving certain indications. For the sake of saving money, a dual-chamber pacemaker with minimum atrioventricular (AV) delay was sometimes used as a biventricular VVI system. At the turn of the millennium, CRT systems still had electrically connected right- and left-ventricular channels. Today, right ventricular (RV) and LV leads may be programmed separately. LV parameters of pacing pulse width and amplitude, pacing configuration, and, to a limited extent, timing [47] may be programmed.

As in conventional dual-chamber pacemakers, various pacing modes can be chosen in CRT systems. The DDD mode is suitable for patients with heart failure and sinus bradycardia because it can deliver biventricular pacing synchronous with the atrium at rates above the lower rate limit (LRL) and AV sequential biventricular pacing at the LRL or a sensor-indicated rate. The VDD mode is suitable for patients with heart failure with normal sinus rhythm; in this mode, the biventricular pacing is delivered synchronously with the atrium, but atrial pacing is excluded. The VDD(R) pacing can be inappropriate, though, because AV desynchronization occurs during sensor-controlled ventricular pacing if the sensor-indicated rate exceeds the intrinsic sinus rhythm. The VVI(R) modes may be harmful to patients with heart failure with normal sinus activity but appropriate for patients with chronic atrial tachycardia. They deliver biventricular pacing at the LRL or at the sensor-indicated rate. If tracking to ventricles occurs during atrial tachycardia, a higher LRL can be programmed or a sensor can be turned on.

To secure biventricular pacing, LV and RV pacing must be applied immediately after a possible sensed RV event. This basic function of CRT systems may be used in the modes of tracked or merely ventricular pacing. The pacing rate is between the LRL and the maximum pacing rate (MPR). In modes with an adaptive rate, or in tracking modes, maximum biventricular pacing is limited by the maximum sensor rate or the maximum tracking rate (MTR).

Biventricular pacing can be interrupted by sensed ventricular events, ventricular extrasystoles, spontaneously tracked atrial tachycardia, or even sensed atrial pacing in ventricles (crosstalk). The interruption of biventricular pacing may lead to symptoms of heart failure, which is why the device, while sensing the intrinsic ventricular activity when still in the AV delay interval or in nontracking pacing modes, immediately triggers ventricular pacing if the MPR is not exceeded.

D. Korpas, *Implantable Cardiac Devices Technology*, DOI 10.1007/978-1-4614-6907-0_11,

11.2 CRT Pacing Configuration

The configuration of LV lead electrodes provides enhanced programmable pacing and sensing capabilities in the LV channel. Proper programming of LV lead configuration may optimize the operation of the CRT system in terms of electrical as well as clinical parameters. Unipolar, bipolar, or extended bipolar configurations are available. The options depend on the features of the CRT device and the polarity of the implanted lead.

In justified cases, the LV channel can be left temporarily or permanently unconnected, for example, if the lead cannot be placed in an optimal manner, if later epicardial implantation is planned, or if the lead gets loosened and dislocated, or if it is decided to leave the lead implanted and connected but unused.

In some manufacturers' systems, the LV lead can have different pacing and sensing configurations, which allows pacing or sensing to be optimized by the selection of the highest sensed signal or the lowest pacing threshold. The LV lead pacing vector can also affect the resulting width of a biventricularly paced QRS complex, which we aim to minimize.

When using a bipolar LV lead, six pacing and sensing configurations can be programmed (Fig. 11.1):

- LV distal electrode ≫ device (unipolar)
- LV proximal electrode ≫ device (unipolar)
- LV distal electrode ≫ LV proximal electrode (bipolar)
- LV proximal electrode ≫ LV distal electrode (bipolar)
- LV proximal electrode ≫ RV proximal electrode (extended bipolar)
- LV distal electrode ≫ RV proximal electrode (extended bipolar)

When using a quadripolar lead, ten pacing configurations can be programmed in this commercially available system. Theoretically, this number could be increased to 14 if unipolar configurations were used.

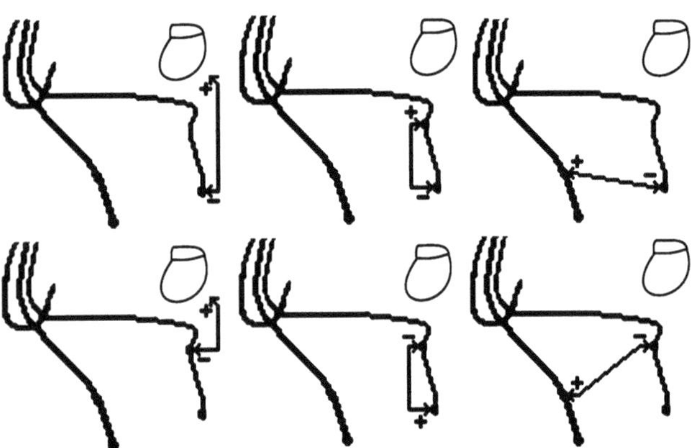

Fig. 11.1 LV lead configurations [82] (© 2012 Boston Scientific Corporation or its affiliates. All rights reserved. Used with permission of Boston Scientific Corporation)

Not all CRT systems have all configurations available. Because of an increased risk of LV lead dislocation, sufficient safety reserve over the pacing thresholds must be programmed for pacing pulses, although it could adversely affect the longevity of the device , which already is reduced by permanent pacing.

11.3 CRT Timing

To ensure optimum CRT delivery, appropriate device parameter values and pacing timing must be programmed. As far as CRT systems are concerned, permanent pacing of the right and left ventricle is considered optimal. Programming the device for separate RV pacing is not intended for the treatment of heart failure; rather, it worsens the situation. LV pacing has not yet been recognized as a clinically valid method to treat heart failure. This must be taken into consideration when setting the timing and parameters. In addition to AV delay, most manufacturers' systems also include interventricular (VV) delay (between RV and LV events in CRT).

If the patient's intrinsic heart rate drops below the LRL, the device delivers pacing pulses at the LRL. If the patient's intrinsic rhythm is lower than the MTR and the programmed AV delay value is lower than the time of intrinsic AV conduction, the device applies pacing pulses in both ventricles according to the programmed setting. To ensure maximum biventricular pacing, the value of AV delay must be set lower than the patient's intrinsic PR interval.

For the purpose of decision making concerning the CRT or bradycardia therapy, a cardiac cycle based on RV sensed or paced events is used and to which all device timing cycles are related. Sensed LV events suppress inappropriate LV pacing and have no influence on the timing cycle. This allows CRT to be applied even in the VVI mode without an atrial lead, for example, in the case of chronic atrial fibrillation.

The parameter of sense AV delay is used to reach a shorter AV delay after sensed atrial events, whereas the programmed value of the pace AV delay parameter is used after paced atrial events. Atrial pacing can prolong the interatrial delay, which is why it can be necessary to program various AV delay parameter settings to optimize CRT during sensed and paced sinus rhythm. Setting optimal AV delay parameters should result in the maximization of the LV filling time. This should be verified after implantation by means of echocardiography or pulse pressure monitoring. The optimized value is often around 100 ms.

Because of the extension of the pacing system with an LV lead, the CRT system is also supplemented with blanking and refractory periods [77, 82]. These periods (Fig. 11.2), however, are not programmable in all manufacturers' systems. For example, the LV refractory period (LVRP) prevents the undesirable loss of CRT after a sensed or paced event, such as a left T wave. Proper programming of this function helps maximize

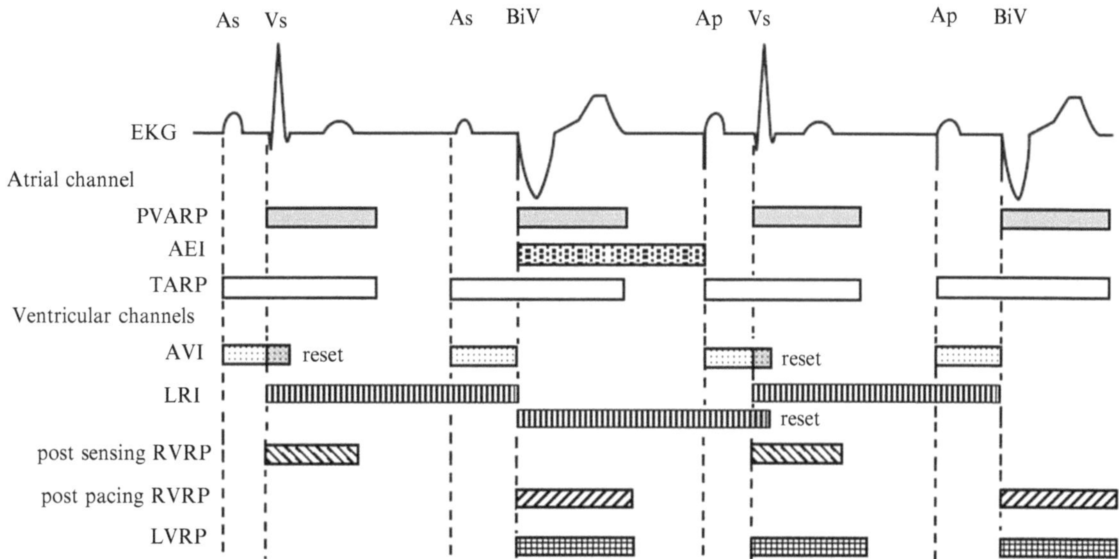

Fig. 11.2 Biventricular pacing timing periods [82] (© 2012 Boston Scientific Corporation or its affiliates. All rights reserved. Used with permission of Boston Scientific Corporation)

the delivery of CRT and simultaneously lowers the risk of acceleration of the patient's rhythm to a ventricular tachycardia. Even though CRT should be delivered permanently, there are certain circumstances in which suppression of therapy is recommended. LVRP is a period after a sensed or paced LV event or after the first paced ventricular event (if a nonzero VV delay is set) during which sensed LV events do not affect pacing timing. The use of a long LVRP shortens the LV detection window. To suppress inappropriate LV pacing, a sufficiently long LVRP is required so that a T wave is included.

The left ventricular protection period (LVPP) is another possible parameter. It is a period after a sensed or paced LV event when the device does not pace the left ventricle. The LVPP prevents the device from delivering pacing during an vulnerable LV phase, for example, if an LV extrasystole occurs. The use of a long LVPP decreases the maximum LV pacing rate and the CRT can be suppressed at higher pacing rates. In patients with heart failure with normal AV conduction, a long intrinsic intracardial AV interval and a long programmed postventricular atrial refractory period (PVARP) may cause the loss of atrial rhythm tracking below the MTR, resulting in inefficient CRT. A PVARP after an extrasystole should be programmed as long as possible because of the risk of pacemaker-mediated tachycardia.

A left ventricle blanking period after atrial pace suppresses LV sensing after atrial pacing. Nevertheless, it is not available and programmable in all manufacturers' devices. Sensed ventricular events in this period are most often caused by far-field sensing from the atrium [77].

LV offset function or VV interval allows difference between the time of application of LV and RV pacing pulses to be set. Thus, it provides flexibility in timing, the advantage of which may be seen when minimizing the width of a paced QRS complex after CRT implantation. Provided that biventricular pacing is close to the MPR, the device adapts the LV offset according to the lowest programmed value of the tachycardia detection zone. The AV delay programmed value is based on RV timing, and it is thus not affected by the LV offset parameter. Depending on the manufacturer, the LV pacing offset and pre-excitation can be set [82].

For the support of the intrinsic atrial rhythm, the LRL parameter should be programmed to a value lower than the patient's resting rhythm; the MPR, however, should be programmed to the highest rate that can be tolerated by a patient to sustain CRT at fast atrial activity. If the patient's atrial activity exceeds the MPR, CRT can be disturbed by the loss of AV synchronization. Nevertheless, the devices usually support the possibility of CRT application during atrial tachycardia. Events sensed in the atrium may come during the PVARP period, and the device thus classifies them as refractory and fails to track them to ventricles. Tracking interruptions can thus restrict CRT application, which is why some algorithms temporarily shorten the PVARP, so that fast atrial events can be tracked to ventricles and the CRT can be sustained. CRT delivery and programmed AV synchronization with the sense AV delay parameter is restored after the restoration of normal sinus rhythm. Naturally, atrial rhythm tracking is suppressed during tachycardia therapies, electrophysiologic testing via the device, and similar system operations.

11.4 Diagnostic Features of CRT Systems

In patients with an implanted CRT system, progression of heart failure or its further development should be assumed after implantation. For this reason, diagnostic functions in the devices for CRT are much more important than in conventional pacemakers and implanted cardioverter-defibrillators.

During routine follow-up, increased attention should be given to the pacing history in the device's memory.

As in other systems, information on the detection and therapy for tachycardia episodes is recorded automatically. Then the data can be searched at various levels of detail using a programmer. The stored data include episode details with electrograms and annotated markers from all active leads. As soon as the memory capacity available for storing episode data is filled, the device attempts to assign priority to types of stored episodes and rewrites the stored episodes according to specific rules. Depending on the type of device and the manufacturer, the trend function can provide, for example, a graphic representation of atrial and ventricular arrhythmias, the trend of a patient's intrinsic heart rate, the rate of a patient's daily activity measured by an accelerometer, the time of atrial tachy response mode switching, and the trend of a patient's daily respiratory activity, as well as data concerning the long-term trends of sensed intrinsic heart rate and impedance amplitudes.

Heart rate variability (HRV) is the measure of changes in a patient's intrinsic heart rate intervals for a specific period and can help evaluate the clinical condition of patients with heart failure. Lower variability of a patient's heart rate can facilitate recognition of the decompensation condition during heart failure. The HRV monitoring function provides information using the data of sensed intrinsic cardiac intervals. The device can gather information on the intervals for HRV only as a tracking mode without an adaptable pacing rate (VDD or DDD) being programmed. Certain special data useful for detailed evaluation are recorded, and only those RR intervals meeting set heart failure data collection criteria are evaluated. The analysis of RR interval variability can be conducted using two methods: either a spectral analysis in the frequency domain or by parameters in the time domain.

Spectral analysis is most often carried out by means of fast Fourier transform, where sensed RR intervals are divided into characteristic frequency bands. They are classified as high (0.15–0.40 Hz), low (0.04–0.15 Hz), very low (0.0033–0.04 Hz), and ultra low (below 0.0033 Hz). Spectrum samples are recorded from different time intervals according to the frequency band examined. The proportion of low-frequency and high-frequency components of heart rhythm spectral analysis is considered the measure of sympathovagal balance and expresses modulation of sympathetic nervous system activity. Parasympathetic nervous system tone is shown primarily in the high-frequency component of the spectral analysis. The low-frequency component is affected by the sympathetic as well as parasympathetic nervous systems. The calculation, which is carried out by devices with this capability based on a measured RR interval, serves as an approximation of the actual low-frequency and high-frequency component proportions.

Time parameters include the calculation of indices having no relation to a specific cardiac cycle value. These indices represent a simple method of identifying patients with a decrease in heart rate variability. It is, for example, a standard deviation of averaged normal R to R intervals in 5-min periods of time, a standard deviation of all normal R to R intervals, a percentage of adjacent RR intervals that vary by more than 50 ms, or the root mean square of the difference between the coupling intervals of adjacent RR intervals. In addition, weighted or geometrical means of RR intervals can be calculated [83].

As mentioned earlier in this chapter, 100 % biventricular pacing must be ensured for good efficiency of CRT systems. Therefore, it is important to monitor the event counters for individual leads and the proportion of paced and sensed events during follow-up. This is expressed as a percentage of atrial, LV, and RV pacing. Optimally, all events in the atrium should be sensed (or controlled by a sensor), and all events in both ventricles should be paced. Each counter shows the total number of paced events since the reset.

12 Implantation, Explantation, and Replacement of Devices and Leads

If any patient meets the indication of treatment by some implantable system, this treatment starts by implanting the system. In the case of a secondary preventive indication, patients who underwent an arrhythmic accident usually stay at a hospital up to the implantation. In the case of a primary preventive indication, patients present at a hospital on the agreed-upon date, in accordance with a waiting list. This chapter describes techniques for implanting leads into various heart chambers and deals with the invasive interventions that relate to the devices themselves.

Before implantation, all the medical devices intended for use are packaged and ready for use in the original sterile containers. The implantation of re-sterilized medical devices is prohibited. The sterile package consists of the external and internal container. The outer package and sterile tray may be opened by an authorized person only under clean conditions. The sterile tray is opened by peeling the cover back; then only a scrubbed, masked, and gowned person is allowed to open the sealed inner sterile packaging using accepted aseptic technique to ensure sterility. The inner tray is also opened by peeling the cover back.

During implantation, the instrumentation for heart monitoring, imaging (fluoroscopy), external defibrillation, and measuring electrical parameters must be available. When using electric instrumentation, always isolate the patient from potentially hazardous leakage currents. Common practice indicates that sterile duplicates of all implantable items should be available for use if accidental damage, failure, or contamination occurs.

Furthermore, some possible reactions upon implantation of the cardiac pacing systems are lister here (in alphabetical order) [49]:

- Air embolism
- Allergic reaction
- Bleeding
- Component failure
- Conducting of current or insulation of myocardium during defibrillation by internal or external leads
- Death
- Erosion/extrusion
- Excessive growth of fibrous tissue
- Formation of hematomas or cysts
- Chronic nerve damage
- Impossibility of therapy application
- Inappropriate or excessive sensing
- Inappropriate therapy
- Infection
- Keloid formation
- Lead abrasion
- Lead displacement or dislodgment
- Lead fracture or lead insulation break
- Deformation or breakage of the lead tip
- Local reaction of tissue
- Loose connection of the lead pin to the device
- Myocardial injury
- Myocardial irritability
- Myocardial perforation/tamponade
- Pacing threshold elevation
- Pneumothorax
- Thromboembolism
- Transvenous lead-related thrombosis
- Venous blockage
- Venous perforation/erosion

12.1 Implanting Endocardial Leads

Endocardial pacing leads might be implanted using one of the following methods:

- Via venous cutdown through the left or right cephalic vein,
- Via subclavian venipuncture or via venous cutdown through the subclavian vein or internal jugular vein.

Only one incision is required to insert the lead through the cephalic vein (under the clavicle). The endocardial lead is inserted into the right or left cephalic vein in the deltopectoral groove. The vein pick packaged with the lead may be used during the venous cutdown procedure to simplify the implantation of the lead into the vein.

D. Korpas, *Implantable Cardiac Devices Technology*, DOI 10.1007/978-1-4614-6907-0_12,

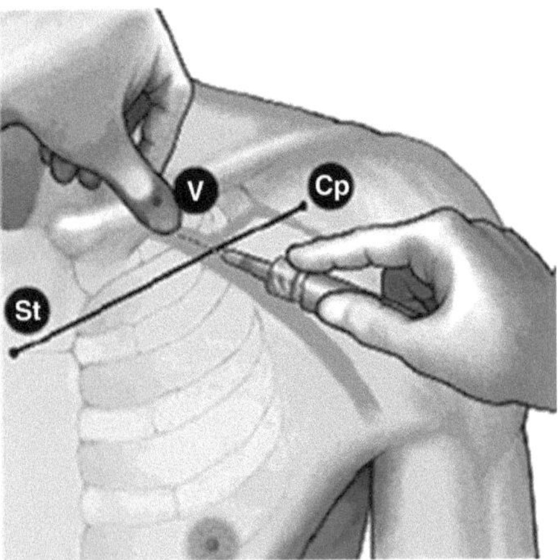

Fig. 12.1 Location of puncture needle entry [49] (© 2012 Boston Scientific Corporation or its affiliates. All rights reserved. Used with permission of Boston Scientific Corporation)

When the lead is inserted using the subclavian venipuncture method, a venipuncture kit and a subclavian introducer are used. The kit contains a venipuncture needle, a syringe, a hemostasis valve introducer, a dilator of the appropriate diameter (in accordance with the lead diameter), and a guide wire. When attempting to implant the lead via the subclavian venipuncture, do not insert the lead under the medial one third of the clavicle because the lead may be damaged. The lead must enter the subclavian vein near the lateral border of the first rib and it must avoid penetrating to the subclavius muscle. Complying with this safety measure is important to ensure that the lead is not damaged by the clavicle or by the first rib. The lead inserted by percutaneous subclavian venipuncture must enter the subclavian vein, where it passes over the first rib to avoid entrapment by the subclavius muscle or ligamentous structures associated with the narrow costoclavicular region. So, it is suggested that the lead be inserted near the lateral border of the first rib. The syringe must be positioned directly above and parallel to the axillary vein to reduce the risk that the needle will contact the axillary or subclavian arteries or the brachial plexus.

The instructions below explain how to identify the venous entry site and define the course of the needle toward the subclavian vein where it crosses the first rib.

- According to Fig. 12.1, identify the sternal angle (St) and coracoid process (Cp).
- Visually draw a line between the St and Cp points and divide this segment into thirds. The needle should pierce the skin at the junction of the middle and lateral thirds, directly above the axillary vein.
- Place your index finger on the junction of the medial and middle thirds (point V), beneath which point the subclavian vein should be located. Compress a thumb against the index finger and project 1 or 2 cm below the clavicle to protect the subclavius muscle from the needle.
- Using the thumb, identify the pressure of the needle as it goes through the surface fasciae; lead the needle to deep tissues in the direction of the subclavian vein and the first rib that lies under it.

If the selected vein is punctured successfully, remove the syringe from the needle while keeping the needle's position in the vein. Pass the guide wire through the cylindrical guiding sheath into the needle and then further through the needle to the vein. Remove the needle and put the dilator on the correctly positioned guide wire that routes caudally through the superior vena cava. The dilator and the introducer are inserted jointly through the guide wire to the vein. When the guide wire is removed, the lead and the stylet are introduced into the required position. After measurement of electric parameters, the lead is fixed. Tear the introducer apart to remove it. In cases of a difficult venipuncture, it is necessary to introduce two endocardial leads from the only successful venipuncture. In this case, leave the guide wire in the introducer and remove only the dilator. During this procedure, use the first venipuncture kit with a respectively bigger diameter. Do not remove the guide wire, and insert the lead and the stylet to the required position. To avoid blood loss and aeroembolism, close the introducer mouth using a finger. After measurements and fixation of the first lead, remove the first introducer. Insert the second dilator and continue using the same procedure as in the case just one lead implantation.

As the lead is implanted by means of the subclavian venipuncture, allow slack in the lead between the distal suture sleeve and the venous entry site. This will help to minimize flexing at the suture sleeve and interaction with the clavicle and first rib region.

12.1.1 Positioning the Lead in the Right Atrium

With the straight stylet in the lead, advance the lead transvenously into the right atrium. When the lead is far enough within the right atrium, introduce the *J*-shaped stylet or a bended straight stylet. Carefully pull the lead and stylet together at the venous entry site to check for contact between the electrode tip and endocardium. A part of the lead sterile package might include various *J*-shaped stylets. A longer action radius is suitable for most patients; bending a smaller stylet might be suitable for patients with a smaller atrium who underwent heart surgery. The most suitable position of the electrode tip is in the right atrial appendage (right auricle) opposite the endocardium (see Fig. 12.2). Use the fluoroscopic anteroposterior projection to check whether the electrode tip is directed medially toward the left atrium.

After the lead tip is fixed to the heart wall, check for the proper motion of the lead. It is necessary to ensure adequate slack of the lead in the atrium that helps to decrease the risk of dislodging the lead. As the patient exhales, the *J*-shaped

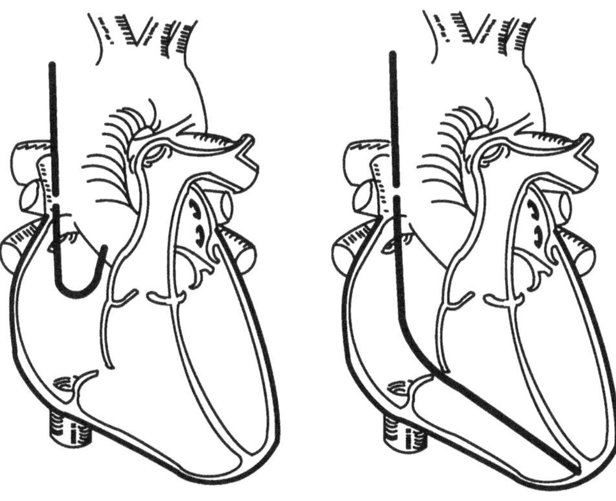

Fig. 12.2 Optimal atrial and right ventricle lead placement [49] (© 2012 Boston Scientific Corporation or its affiliates. All rights reserved. Used with permission of Boston Scientific Corporation)

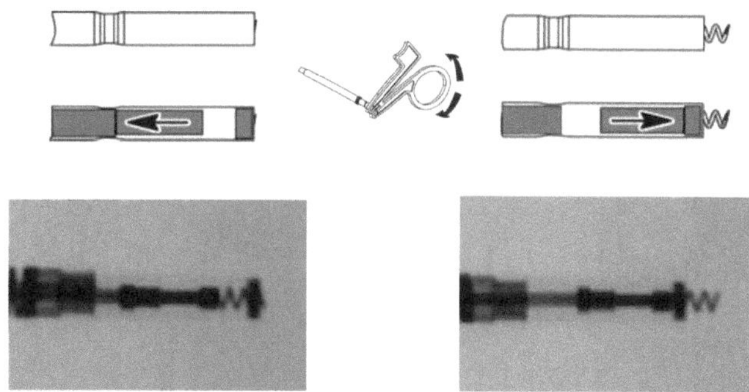

Fig. 12.3 Views of the active helix electrode [49] (© 2012 Boston Scientific Corporation or its affiliates. All rights reserved. Used with permission of Boston Scientific Corporation)

lead must be positioned secure in the atrial appendage. As the patient inhales, the *J*-shaped lead must straighten (it forms an *L* shape). If the lead moves closer to the tricuspid valve during breathing, the slack is excessive [49, 50].

12.1.2 Positioning the Lead in the Right Ventricle

With the straight stylet in the lead, advance the lead transvenously into the right atrium. Further advance the lead through the tricuspid valve or position the lead tip opposite the lateral atrial wall and withdraw the lead body back through the tricuspid valve. Maneuverability might be improved by bending the stylet. By means of fluoroscopy projection, check to ensure the lead did not get stuck in the coronary sinus and that it is really positioned in the ventricle. Introduce the straight stylet into the lead and carefully pull the lead and stylet together at the venous entry site to check for contact between the electrode tip and the endocardium (Fig. 12.2). We recommend positioning the tip of the right ventricular pacing lead at the right ventricle at the mid-septum, respective to the apex of the right ventricle.

12.1.3 Fixating the Active Fixation Lead

The helix of the active fixation lead is electrically conductive so that it is possible to map considered positions of the electrode. The mapping means measurement of pacing and sensing characteristics without entrapping the helix in the tissue. If the values are acceptable, continue with the procedure of the lead fixation. We recommend execution of mapping in the atrium or ventricle before the fixing the lead because repeated relocation of the lead can be avoided. As the lead is fixed or repositioned, the stylet must be extended completely. When the proper position is achieved, connect the fixation tool to the connector pin. Press the handle to position the pin in the groove and release the handle to block the pin in the fixation tool.

Lodge the distal electrode at the required place using adequate tension. Rotate the fixation tool clockwise to extend the helix so that the distal end of the lead is fixed to the cardiac wall. Usually 8 to 10 turns at a rate of about 1 turn/s are sufficient to ensure helix penetration and to transfer the torque. Observe the fixed helix extension on the fluoroscopy monitor with X-ray contrast markers and visually check its position (Fig. 12.3).

Excessive rotation might damage the lead, increase the acute effects of the voltage thresholds, dislodge the electrode, or cause lead perforation. When the fixation tool is released, the connector pin starts antirotation. Press the tool handle and remove the fixation tool from the connector pin and carefully remove the stylet. Using fluoroscopy, check whether the lead has sufficient slack, which decreases risk of its dislodgement. In a case of dislodgement of the lead, immediate intervention is necessary because the electrode position must be reestablished as soon as possible and the damage on the endocardium must be minimized.

If repositioning the lead is necessary, introduce the stylet fully again, connect the fixation tool, and rotate the fixation tool counterclockwise until the helix is fully retracted. Again, excessive rotation might damage the lead. Using fluoroscopy, check whether the helix is fully retracted and completely dislodged from the heart wall. In the case of eventual lead extraction, parallel counterclockwise rotation by the lead body is necessary to avoid spontaneous damage of the tissue. Counterclockwise rotation of the lead also helps to avoid accidental fixation.

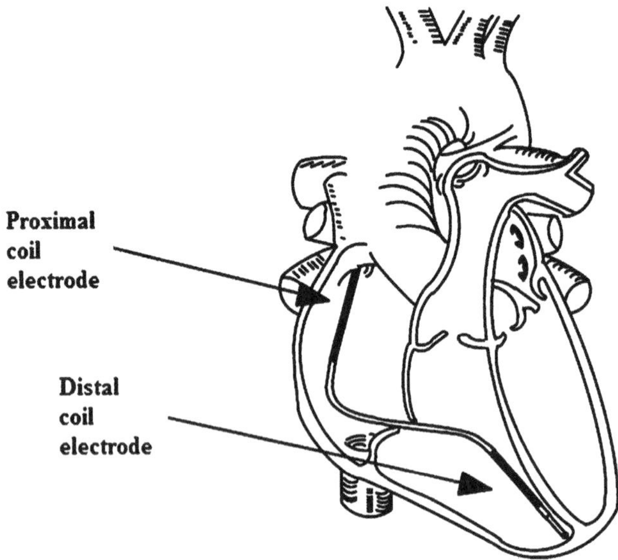

Fig. 12.4 Optimal position of the defibrillation lead in the right ventricle [51] (© 2012 Boston Scientific Corporation or its affiliates. All rights reserved. Used with permission of Boston Scientific Corporation)

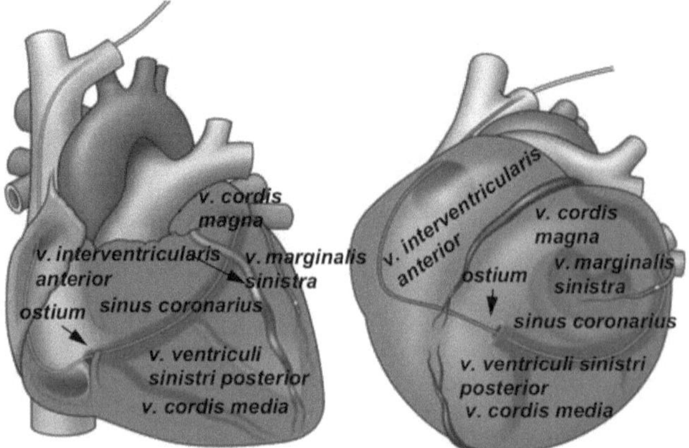

Fig. 12.5 Anteroposterior and left anterior oblique views of the coronary venous system [82] (© 2012 Boston Scientific Corporation or its affiliates. All rights reserved. Used with permission of Boston Scientific Corporation)

12.1.4 Fixating the Passive Fixation Lead

The lead with the stylet introduced is advanced transvenously to a place where it can be captured by the trabeculae. Where the lead tines get embedded, the stylet is partially retracted. Using fluoroscopy, check the stability of the lead in the trabeculae as the patient coughs and respires deeply on your upon the clinician's instruction. If it is necessary to change the lead position, advance the lead's pacing tine to the trabeculae by means of a straight stylet. It is necessary to pay attention to the fact that all the leads must be positioned in the healthiest heart tissue available.

12.1.5 Implanting the Defibrillation Lead

Apply the defibrillation lead using fluoroscopy; the stylet is introduced to the lead so that its distal end is placed in the right ventricular apex. Check whether the distal shock electrode is positioned in the right ventricle under the tricuspid valve (right ventricular apex) and that the respective proximal electrode (in case of dual-coil defibrillation leads) is situated at the superior vena cava and at the right atrial junction. Correct function of the leads is dependent on their proper position. The defibrillation lead also serves for pacing, so its distal tip must be situated at the healthy myocardium in the apex of the heart (Fig. 12.4). Improper positioning might cause movement of the lead and prevent the defibrillation shock from incorporating the apex of the heart.

12.1.6 Implanting the Left Ventricular Lead

The left ventricular (LV) lead is positioned transvenously by means of catheterization made by a long guiding sheath and a mapping electrophysiological catheter. First, cannulate the ostium of the coronary sinus. The diameter of the mapping electrophysiological catheter is usually 6 F (or smaller), and its distal end is equipped with a deflectable or flexible tip. After removing the mapping electrophysiological catheter, the route for lead positioning is formed. Take the coronary angiogram with or without the help of a balloon occlusion catheter. The angiogram visualizes the system of coronary veins. Save the angiogram for later reference when dealing with the venous anatomy. Risks connected to this intervention are similar to those of any other catheterization procedure of the coronary sinus. Some patients might show intolerance to various types of contrast media. Figure 12.5 displays an example of the coronary venous system. The coronary sinus and its branches include the great cardiac vein (*v. cordis magna*), the middle cardiac vein (*v. cordis media*), the posterior vein of the left ventricle (*v. ventriculi sinistri posterior*), and the left marginal vein (*v. marginalis sinistra*). Different anatomic conditions among patients enable positioning of the lead in one or more recommended places.

The guiding sheath helps to introduce the lead into the venous system and helps to protect the LV lead during positioning of the other leads. To avoid thromboembolism in the lead and the guiding sheath, flush the internal lumen of the lead and the guiding sheath with heparinized physiological solution before and during their usage. Introduce a suitable guide wire to the lead through the distal electrode tip so that the wire is projected less than 2 cm and check whether the guide wire can be pushed easily through the lead's lumen.

Subsequently introduce the lead through the guiding sheath into the coronary venous system. The lead might also be introduced into the guiding sheath without the guide wire, and this wire can be used later. Some LV leads might be introduced by the stylet that changes the shape and flexibility of the appropriate preformed distal end of the lead. Some introducer systems contain another internal catheter that is introduced from the external guiding sheath. This catheter serves for mapping of the most suitable coronary vein for implantation of the lead.

After the lead is positioned in the required place, the tip of the guide wire is partially retracted to the pacing lead so that it does not overhang the lead tip. When the lead is positioned correctly, the guide wire is removed. Then, remove the finishing wire from the package and insert it into the lead according to the manufacturer's instructions. Both the lead and finishing wire are kept at the position required for implantation and, at the same time and in accordance with the type of sheath, the guiding sheath is removed by cutting or drawing over. If preformed leads are used, do not advance the finishing wire at a full stretch but leave it drawn out lightly when the guiding sheath is being removed. When the preformed lead end is applied that way, it ensures higher stability. After the guiding sheath is removed, check again to ensure the distal end of the lead did not change position. The proximal end of the lead is kept close to the venous entry site. The finishing wire is disconnected from the connector by rotation and retraction and then is removed from the lead. Using fluoroscopy, check again to ensure the lead did not move. The finishing wire must not be stuck in the lead. If the finishing wire cannot be retracted from the lead, remove the finishing wire and the lead as one unit [56, 57].

12.1.7 Securing the Lead at the Venous Entry Site

After the lead is positioned satisfactorily and fixed in the endocardium, secure the lead to the vein using the suture sleeve provided. Securing the lead will provide permanent hemostasis and lead stabilization. The suture sleeve tie-down technique can vary with the technique used to insert the lead.

For the venous cutdown method, slide the suture sleeve into the vein past the distal preformed groove. Ligate the vein around the suture sleeve to obtain hemostasis. Use the proximal preformed groove to secure the suture sleeve to the lead. By means of the same groove, secure the suture sleeve and the lead to the adjacent fascia. After the lead is secured, check to ensure that the lead does not slip in either direction (Fig. 12.6).

For the subclavian venipuncture method, peel back the introducer sheath and slide the suture sleeve deep into the tissue (Fig. 12.6). Use the grooves to ligate the suture sleeve to the lead and secure the sleeve and lead to the fascia. Then check again to ensure the lead does not slip in either direction.

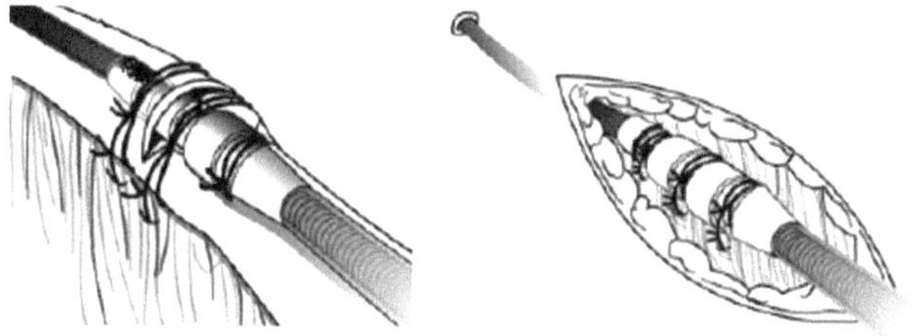

Fig. 12.6 Using a suture sleeve with a percutaneous implant [49] (© 2012 Boston Scientific Corporation or its affiliates. All rights reserved. Used with permission of Boston Scientific Corporation)

12.2 Evaluating Electrical Performance of Pacing Leads

Before attaching the lead to the device, perform a test of lead's electrical parameters. Check these parameters roughly before fixing the lead and again after fixing and suturing the lead. For the measurements, pacing system analyzers (PSA) are used. Connect the lead connectors to a PSA recommended by manufacturer. The connector pin of the bipolar leads forms the cathode (−). The pin must be connected to the negative conductor of the PSA's patient cable. The ring of the lead connector forms the anode (+). Connect it to the positive conductor of the patient's cable. The sensed signals can also be measured by an electrocardiographic recorder or an oscilloscope. Depending on the possibilities of the device, several pacing configurations might be measured at the left ventricular leads.

Electric performance is evaluated by monitoring the data described in the following sections.

12.2.1 Amplitude of Intracardial Signal Sensing

Sufficient amplitude of the intrinsic intracardial electric signals (P wave in the right atrium, R wave in the ventricles, or both) expressed in millivolts is necessary for permanent sensing of the implantable device by the input sensing amplifier. No pacing is required during this measurement. The purpose is to assure that the intrinsic intracardial signal is of adequate amplitude where it is being detected by the device and to inhibit pacing. Furthermore, the integrity of the sensing circuit must be assured. It deals with the integrity of the lead and the lead-to-tissue touch. The intrinsic implantable devices might measure baseline-to-peak amplitude

Fig. 12.7 Acute and chronic electrogram record

while the PSA might measure peak-to-peak amplitude or vice versa. Various bandpass filters in implantable devices and PSAs attenuate signals that are not typical of P wave and R wave frequency. For these reasons, the amplitudes of the intrinsic intracardial electric signals measured first by the PSA and then by the device in situ might differ a lot. The electrogram taken at the time of implantation will show ST segment elevation due to local myocardial damage (see Fig. 12.7). The long-term electrogram will show no further S-T elevations and lower amplitude of the R wave.

12.2.2 Pacing Circuit Impedance

Impedance deals with the total resistance to the flow of electric current through the lead's conductors, electrodes, lead-to-tissue interface, and body fluids and tissues. It is measured in Ohms. This measurement requires delivery of a pacing pulse. The purpose is to verify the integrity of the pacing circuit at the given position of the lead. Over the long term, the decrease in impedance might indicate problems with the lead's insulation, whereas an increase in impedance might indicate problems with the lead's conductor, lead-to-tissue touch, or contact of the lead's connector in the device header.

12.2.3 Pacing Threshold (Voltage/Current)

The pacing threshold is a minimal value of electric pacing that consistently produces cardiac depolarization. It is stated as voltage amplitude of rectangular pulse at a defined pulse width (most often it is 0.4 or 0.5 ms) or, rarely, as current amplitude or pulse width of a given voltage. The purpose is to ensure an adequate safety margin between the pacing threshold and the programmed output of the device at the given position of lead. Furthermore, it is necessary to assure the integrity of the pacing circuit. The threshold is measured by a gradual decrease of the pacing voltage/current/pulse width at an adequate pacing rate. The last value before the loss of pacing is the pacing threshold.

Presence of the electrode next to the endocardial surface causes an inflammatory reaction that heals within several months, leaving behind a small fibrotic capsule around the electrode. After implantation, set the output amplitude and the width of pacing pulse to the device's nominal values; at a minimum this should be double the value of the measured voltage pacing threshold. For the permanent setting, double the value of the measured voltage pacing threshold or triple the value of the pulse width at the threshold amplitude is recommended.

12.2.4 Slew Rate

The slew rate parameter shows behavior of the intracardial signal voltage in time. It deals with the inclination of the leading edge of the intracardial signal impulse. It is measured in volts per second or millivolts per millisecond. In general, the higher the slew rate is, the higher the frequency of the signal and the more likely it is to be sensed. It may be used as an adjunct measurement if the intracardial signal amplitude is on the borderline of acceptable/unacceptable. The purpose is to ascertain whether the sensing is trouble free. If the slew rate is low (ventricular slew rate < 0.5 V/s or atrial slew rate < 0.3 V/s), the frequency is also low and therefore a larger amplitude is needed before the signal can be sensed. Signals with a low slew rate can potentially result in undersensing.

12.2.5 Retrograde Conduction Time

This parameter deals with the period of time it takes an electric impulse originating in the ventricles to travel through the conduction system to the atrium, where it causes atrial depolarization (retrograde P wave). Retrograde conduction may result in endless-loop tachycardia in the atrial-tracking pacing modes when the retrograde P wave is sensed by the atrial sensing circuit and begins an atrioventricular delay. This testing may be deferred until the first follow-up after implantation and event markers from the programmer may be used for the monitoring. Optimal programming of the Postventricular atrial refractory period parameter to the value at least 25 ms longer than the measured retrograde conduction time serves for the prevention of pacemaker-mediated tachycardia. The retrograde conduction time may vary from 100 to 400 ms.

12.2.6 Diaphragm Pacing

Measurement of diaphragm pacing is performed by temporary pacing at the maximal amplitude (10 V). The purpose is to verify that phrenic nerve stimulation (in the case of an atrial lead) and diaphragm stimulation (in the case of ventricular leads) does not occur. This might cause hiccoughing

Table 12.1 Recommended electrical performance

	Implant	Chronic
Sensing amplitude		
Atrium	≥ 1 mV	≥ 1 mV
Ventricle	≥ 5 mV	≥ 5 mV
Impedance		
Pacing circuit	300–2,000 Ω	300–2,000 Ω
Shock circuit	20–80 Ω	20–80 Ω
Pacing threshold		
Voltage (at 0.5 ms)		
Atrium	≤ 1.0 V	≤ 1.5 V
Ventricle	≤ 1.0 V	≤ 1.5 V
Current (at 0.5 ms)		
Atrium	≤ 1.5 mA	≤ 6 mA
Ventricle	≤ 1.5 mA	≤ 6 mA
Slew rate		
Atrium	≥ 0.3 V/s (mV/ms)	≥ 0.3 V/s (mV/ms)
Ventricle	≥ 1.0 V/s	≥ 0.5 V/s
Retrograde conduction time	100–400 ms	

or pectoral muscle stimulation in the pocket where the device is implanted.

Later measurements by means of the implanted devices needn't strictly correspond to the measurements established by means of the PSA as a consequence of the signal filtering. However, the basic measurement should correspond to the recommended values stated in Table 12.1.

Electrical performance of the lead should be measured after allowing sufficient time for the effect of local tissue trauma to subside (about 10 min). If the measurements do not correspond to the values stated above, it is necessary to reposition the lead and secure it again using the procedures previously discussed.

The initial electrical measurements may differ from the long-term recommendation because of acute cellular trauma. The values might be dependent on patient-specific factors, such as the state of tissue, the balance of electrolytes, and medicamentous interactions. Lower internal potentials and higher pacing thresholds might be a marker of the lead position in the ischemic or cicatricial tissue. Because the signal quality might worsen over time, it is necessary to change the lead position so that the obtained signal has the highest possible amplitude, the shortest possible period, and the lowest possible pacing threshold. In the permanent state, R wave amplitudes lower than the recommended value might cause incorrect determination of heart action rate, which might lead to failure of tachycardia detection or a false-positive diagnosis of normal rhythm. In addition, a low value of the pacing threshold requires a safety limit because the pacing threshold might increase after implantation. Signal periods that exceed the programmed refractory period of the device might cause incorrect determination of the cardiac action rate, unsuitable delivery of the high-voltage shock, or both.

12.2.7 Interrogation and Testing of Implanted Device

Interrogate the device by means of the programmer before opening its sterile package. To ensure the correct measurement of parameters, heat the device to room temperature. Perform tests of the pacing or tachycardiac modes, form the condenser manually, check the state of the battery, and reset the counters – all of this depends on the type of device. If the actual state of the battery indicates something other that "beginning of life," the device must not be implanted. Defibrillators also indicate so-called monitored battery voltage that must be at the maximal level determined by the manufacturer before implantation. Before connecting the leads, especially in patients who are dependent on cardiac pacing, the device must be programmed and switched over from the storage mode to ensure immediate pacing after the connection.

12.3 Connecting Leads to the Device

Insert the lead connector straight into the lead port without bending the lead near the lead–header interface. Incorrect insertion might cause damage to the lead's insulation. If required, or in accordance with manufacturer's instructions, lubricate the lead connector sparingly using sterile water or gel to make insertion easier. If the lead connector pin is not connected to the device at the time of implantation, the lead connector must be capped before closing the pocket incision. Tie down a suture around the lead cap to keep it in place. With regard to the patient's anatomical structure and the size and movement of the device, gently coil the excess lead wire and place it in the pocket mostly under the device. It is important to place the lead into the pocket in a manner that minimizes tension and torsion, the cranking to the closed angle, and reduces pressure.

Implantable systems are equipped with various types of lead connectors. IS-1 connectors are most common for cardiac pacing leads. Before inserting a lead's connector to the device header port, it is necessary to check whether the setscrews are extended enough to enable insertion of the connector. If the IS-1 connector is connected to the device correctly, the pin must visibly project at least 1 mm behind the connector block. For correct connection of the older 3.2-mm connector (different than the IS-1 type), the pin must reach less than 1 mm from the socket end after inserting the lead. Then, insert the torque wrench into the slot in the center of the pothole in the seal plug positioned near the tip of the completely extended lead's connector. The device is equipped with setscrews that are to be tightened using a bidirectional torque wrench that is a part of the supplied package. While tightening the setscrews, push the lead so that it stays completely inserted in the socket.

The setscrews should be tightened until the wrench starts skipping. Further tightening is needless because the torque wrench is set so that it applies the proper force on the setscrew. In the case of device replacement, an adaptor that enables connection of a new device to the existing leads might be needed. Information on available adaptors is also stated in Clause 7.8.

After interconnecting the connectors, position the device into the pocket under the skin or muscle. If dealing with unipolar pacing, it is necessary to ensure creation of permanent contact between the device and the tissue, otherwise the device will not operate correctly. Before suturing, verify the device function on the electrocardiogram monitor. Suture the device to the tissue by means of the eyelets usually placed in the upper part of the device header. After the implantation of a new device where a bigger device was placed previously, air bubbles might be captured in the pocket, the device might move or erode, or insufficient contact might occur at the device –tissue interface. The probability of capturing air or of insufficient contact might be reduced by filling the pocket with a physiological solution. The possibility of movement or erosion is reduced by stitching the implant with a fixed suture. Again check the correct function of the device and correct tightening of the setscrews before closing the pocket.

12.4 Device Replacements

The device might need to be replaced for various reasons. The most common reasons for replacement are the end of battery life or infection of the pocket. Further reasons include an upgrade to a dual-chambered or biventricular system or technical problems with the device [84].

To a great extent, the success of the procedure is dependent on the preoperative planning and technique. Before the intervention it is necessary to ensure that all necessary equipment is available for the intervention in case of complications. The equipment for monitoring and display, temporary pacing, and external defibrillation of the patient must be in place. When the patient is connected to the monitor, the defibrillation therapy of the device must be switched off to avoid unnecessary shocks. The electric signals that come to the body from some external monitoring devices or as the setscrews are loosened might cause high rate pacing. Consider switching off all functions that influence the pacing rate because the sensor's response might be activated by handling the device as it is being removed from the pocket. Electrosurgical cautery at the leads or in their immediate surroundings might cause ventricular arrhythmias or fibrillation. Using electrocauterization at the device or in its close proximity might be the cause of temporary incorrect functioning of the device.

Apart from other things, the preoperative planning of the replacement procedure includes verifying data dealing with the implant and determining the system parts that are used (including those that are implanted but not used anymore) according to their manufacturer, model, and serial number. As the new device is programmed, consider the mode that was programmed in the old device and diagnostic data that were collected by the old device. Before disconnecting the lead(s), consider the uniqueness of every system. Patients who are dependent on pacing will have temporary pacing delivered so that continuous pacing was ensured during the intervention. In the case of a unipolar configuration, the patient would not have any pacing pulses delivered after the device was removed from the pocket. If the selected replacement device is not compatible with the present leads, an adaptor must be used or the leads must be replaced. Using incompatible leads might cause damage to the connector or it might cause potential adverse consequences such as inadequate sensing of heart's activity or failure to apply the needed therapy. Before exposing the current device, verify the position of the lead using fluoroscopy. Then proceed carefully to avoid perforating the lead's insulation or damaging anything during system disassembly.

After incising the pocket, remove the device carefully so that the leads are not stressed. Change of color on the device's surface is a consequence of the normal anodic reaction and it does not influence the device's function. Slacken the setscrews by means of a calibrated torque wrench. Gently insert the wrench bit to the central pothole of the seal plug at an angle less than 90° to unbolt the screw. Apply light downward pressure until the wrench bit gets completely engaged in the hexagonal pothole in the screw. When the wrench bit is completely engaged, rotate the wrench slowly counterclockwise until the lead connection is dislodged. If the wrench clicks while being rotated counterclockwise, stop the movement – this means the screw is at the end position. When the screw is dislodged, catch hold of the lead as close to the connector as possible and pull it out using a light force. Repeat this procedure for all the leads connected to the device. In most cases of leads replacement, the leads put up minimal resistance as they are removed from the header. Occasionally, the leads might get stuck in the device header, which can happen for many reasons (e.g., stuck setscrews, fluid in the header, some glue left over the lead sheath, conglutination of the lead and header seal plugs, or a small mechanical tolerance of the header–lead interface). Recommendations for how to release the stuck leads are described later.

Check the integrity of the permanently implanted leads at every replacement of the device. Assess the lead's state using X-rays and by visually checking the naked insulation of the lead body, connector, and seal plugs. Using the PSA, check the electric function of the lead after its disconnection from the device. Measure the pacing thresholds, intrinsic amplitudes, and pacing impedances and check whether there is any noise disturbances at every lead. Insert the lead to the new device header after the visual check of the setscrews, which must be sufficiently extended from the device.

If necessary, lubricate the lead connector sparingly using sterile water to make its insertion easier. After tightening the screws, insert the new device to the implantation pocket. If the device is coated, rotate the uncoated opening off the muscle and toward the skin to restrict the possibility of muscle stimulation. In the case of unipolar pacing, the device case must be in electrical contact with the tissue under the skin. Otherwise, pacing must not be delivered to the patient. Before suturing the pocket, interrogate the device by means of the programmer to verify the telemetric communication. If the correct pacing and/or sensing do not work, disconnect the lead from the device and check the connector and leads visually. Inappropriate signals might indicate misalignment of the lead. In such a case it is necessary to correct the lead's position.

If any unexpected complications occur, the leads must be solved first. Several procedures were developed to loosen stuck leads successfully, including:

- Loosening stuck setscrews,
- Lubricating the lead–device interface, and
- Exposing the lead's connector by cutting off the header [84].

These procedures can be used so that the lead does not have to be replaced. The lead may be damaged during any replacement of the device. As with every other intervention, comply with standard procedures and ensure continuous pacing of patients who are dependent on pacing during any process in which the system of heart rhythm control is being removed. At the same time, it is recommended that the integrity of the leads be tested to ensure that the permanent leads are still suitable for use.

12.4.1 Loosening Stuck Setscrews

From time to time, the setscrews might be stuck in the tightened position, which ensures that the leads cannot be retracted from the header. The following procedure enables straightening of the screw mechanism and loosening of the stuck screws by means of a torque wrench.

- Gently insert the bidirectional hexagonal torque wrench into the hexagonal opening in the setscrew through the central pothole in the seal plug. Using a torque wrench different than a hexagonal one might lead to the rounding of the hexagonal opening in the screw.
- In accordance with the direction of the torque wrench rotation, find out whether the screw is stuck in the tightened position.
- Align the screw mechanism so that the stuck screw might be loosened. Incline the torque wrench from the upright position to the side, approximately 20°–30° from the vertical axis of the central screw (see Fig. 12.8). The wrench bit is bent slightly by the inclination of the hexagonal wrench at this angle. Start rotating the wrench in the appropriate direction to retract the screw. This movement looks like the so-called precession motion.
- When the setscrew is loosened, continue its retraction by rotating as necessary.

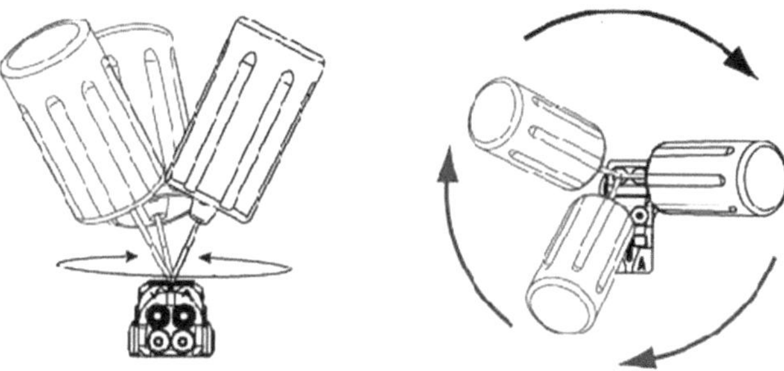

Fig. 12.8 Views of wrench rotation to free stuck setscrews [84] (© 2012 Boston Scientific Corporation or its affiliates. All rights reserved. Used with permission of Boston Scientific Corporation)

The torque wrench that is used for this procedure must not be sterilized and used again because its calibration cannot be ensured.

12.4.2 Lubrication of Lead-Device Interface

If the screws are completely tightened and the lead cannot be removed, it might be stuck because it is clogged by body fluids or a silicone binding substance. In such cases, the terminals might be released by lubrication of the lead–device interface using a sterile heparinized physiological solution.

- Fill a 1-ml syringe with a needle with heparinized physiological solution.
- Gently insert the needle into the device header parallel with the lead between the body of the stuck lead and the seals of the header up to the outermost connector block. The needle must be visible through the header material.
- Inject the heparinized physiological solution to the block of connectors until it starts spilling out from the setscrew cap on the same side of the header.
- Take hold of the lead as close to the connector as possible and pull it out using a gentle force.

During this procedure, the lead's insulation might be damaged, which will cause an immediate or gradual penetration of fluid. After this procedure is used, we recommend planning suitable subsequent checks apart from the careful visual check and thorough testing of electric functions by means of the new device.

12.4.3 Exposing the Lead Connector by Cutting Off the Header

If the methods described above fail, there is the possibility of preserving the leads at the price of the destruction of the implanted device. There is a method that allows a back part of the device header to be cut off and to save the stuck leads.

- Protect the operated anatomy with a suitable cover so that any chippings do not get into the patient's body.
- Hold the device securely with flat-nosed pliers.
- Place the large cutting pliers with a joint on the back part of the header a secure distance from the end of the lead connector but still on the lead so the lead terminal connectors are not damaged.
- Press the handle of the pliers and cut through the lead case in the back part of the header and uncover the connectors' pins.
- When the lead case is uncovered, push the connector pin from the header by means of a torque wrench or pliers and at the same time gently pull the lead connector.

Because the electronic circuits are placed in the area of the cut, cardiac pacing will be aborted by cutting off the header. Therefore, temporary pacing must be ensured for pacing-dependent patients.

12.5 Explantation

12.5.1 Explanting Leads

Explanation or replacement of the endocardial leads are interventions that carry huge risk. This risk increases with the time that has passed after implantation. It is stated that leads might be extracted easily in the first 3 months after implantation [40]. One year or more after implantation, a lot of fibrous tissue bridges along the whole length of the lead, especially in the right ventricle and tricuspid valve and at the location of endocardial fixation. The intervention does not describe the explantation procedure as the reverse order of the implantation; instead it deals with a complex intervention of the lead extraction with the appropriate risks.

12.5.2 Explanting Devices

The technique for device explantation is similar to that of their replacement. Devices and leads explanted ahead of schedule (source discharge, infection, technical problems, and patient death) should be returned to the manufacturer because their investigation might provide information for improvement of the reliability of the device and provision of guarantees. Explanted medical devices must not be implanted to another patient because their sterility, functionality, and reliability cannot be guaranteed. For ethical and technical reasons it is recommended that devices be explanted postmortem.

Before explanting, cleaning, or sending a device, perform the following steps:

- Switch off the pacing and tachycardia mode of the device.
- Switch off the function of magnet response.
- Switch off acoustic end-of-life signalization.
- Print a message about the final programmed mode and states of counters and memory.

Clean and disinfect the device and leads with disinfectant while complying with the techniques valid for biological risk so that body fluids and tissue rests are removed. However, never bathe the device or let fluid penetrate the lead ports. Then, pack the device thoroughly in the return kit and send it to the manufacturer.

13 Patient Follow-Up

Throughout the longevity of a device, both the technical condition of the implanted system and clinical indicators recorded in the device memory in the period between two follow-ups must be monitored on a regular basis. It is an established practice that the initial system check is carried out immediately after implantation; the next follow-up should precede a hospital discharge. Further follow-up should be performed as required by the implantation center either in a month and then in 6-month intervals, or in 3–6 months. Some manufacturers may recommend conducting follow-ups at an interval of 3 months, in particular during the period after implantation. As a rule, a patient is supposed to have the system follow-up after a defibrillation shock occurs, sound signals are emitted by the device, or in other unexpected circumstances.

During implantable system follow-ups, at a minimum an external defibrillator and an electrocardiogram (ECG) recorder or a monitor should be available. For the purpose of optimizing medical personnel, commissioning by a medical care provider or a medical device supplier of a qualified biomedical or clinical engineer to perform routine follow-ups has proved to be convenient. These engineers are greatly experienced in the control of programmers and applications, which shortens the follow-ups and enhances safety. Provided that an electrogram analysis is required, physician/technician consultation is recommended.

At the follow-up preceding hospital discharge (usually within 72 h after implantation), the device should be interrogated using a programmer, and a simplified follow-up should be carried out using a summary screen, which is available in most applications. The patient must be checked for lead dislodgement or other postoperative complications. All leads must be tested for pacing thresholds and lead impedance, and the intrinsic signal amplitude must be measured. In addition, final setup of the system parameters must be performed according to the actual state, and therapy counters must be reset so that only data relating to the latest episodes are displayed at the next follow-up. Then, a record is printed from the programmer and included in the patient's file.

In addition to routine measurement of electrical parameters (pacing thresholds, intrinsic signal amplitudes, and impedances) in all leads and check battery status and stored electrograms, at the first follow-up after discharge it is necessary to verify the appropriateness of the parameters set during the postimplantation phase for patient's daily life and the condition of the implantation wound. To optimize the settings, information concerning the patient's feelings during the postimplantation period (quality of sleep, palpitation) is important. With pacemakers, this primarily involves setting the lower rate limit and parameters of atrioventricular (AV) delay or the sensor. Pacing of the right ventricle (dysfunction prevention) and the right atrium should be minimized to prevent unnecessarily fast pacing. To support the intrinsic heart rate, it is always helpful to set pace hysteresis and AV delay hysteresis, if possible. An AV delay that is too short results in excessive right ventricular pacing and insufficient ventricular filling phase time. In implanted cardioverter-defibrillators (ICDs), the detection zone setting must be rechecked, preferably based on histograms of the intrinsic heart rate, so that the zone lower limit is not close to the possible normal intrinsic rhythm. In addition, any electrograms stored in the device memory must be evaluated.

At further follow-up, electrical parameters of all leads must be measured, the battery status has to be checked, and stored tachycardia electrograms should be evaluated. In patients with a chronic condition, attention must be given to sudden variations in system parameters, such as pacing or a decrease or increase in the shock circuit impedance. Specific diagnostic tools depend then on the respective device features. For the sake of clarity, it is recommended that bradycardia counters and stored tachycardia episodes be reset at every follow-up; the memory is divided into short-term

D. Korpas, *Implantable Cardiac Devices Technology*, DOI 10.1007/978-1-4614-6907-0_13,

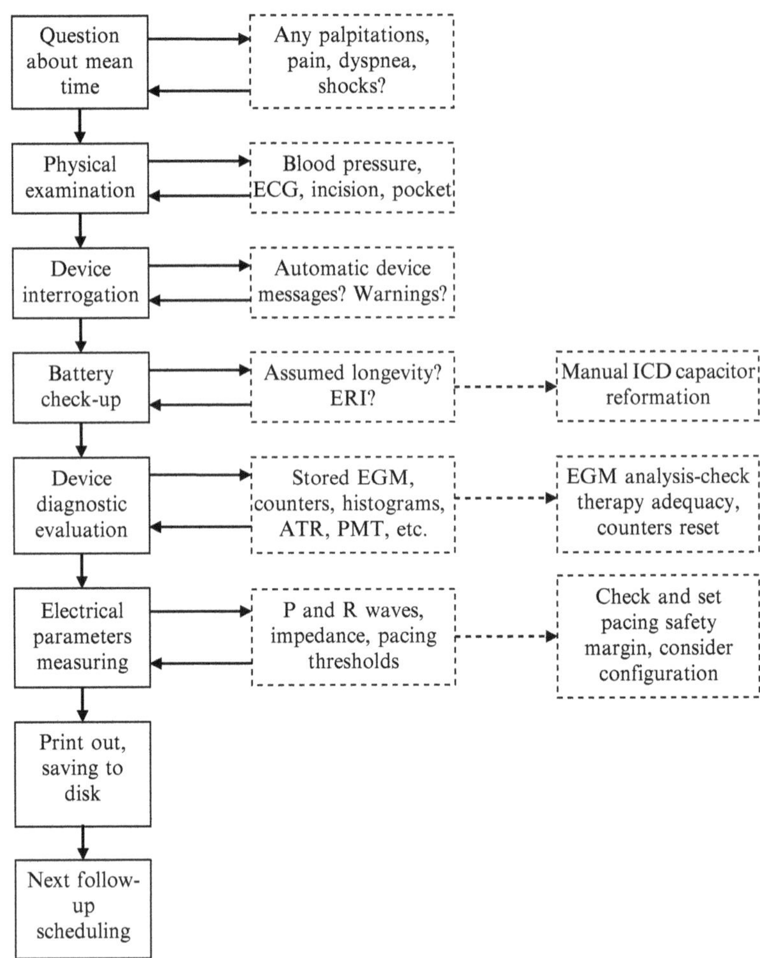

Fig. 13.1 Flow chart for ambulatory device follow-ups

memory (since the previous follow-up) and long-term memory (since implantation). Overall, the follow-up procedure may be described as shown in Fig. 13.1.

The range of electrical parameter values follows from the manufacturer's recommendations. The usual minimum value of intrinsic activity is 1–2 mV for the P wave and 5 mV for the R wave. Pacing impedance usually ranges from 300 to 1,200 Ω; the value for the left ventricular lead can be even higher – up to 1,800 Ω. Depending on the lead type, impedance of the defibrillation circuit reaches approximately 35–60 Ω; provided that only a lead distal shock electrode is used, it is usually higher – up to 90 Ω. The optimum voltage pacing threshold is below 1.0 V at a pacing pulse width of 0.5 ms. The pacing threshold slightly increases after implantation and then decreases again. This phenomenon is suppressed by steroids. Long-term monitoring should prove all the electrical parameters to be more or less on the same level throughout the system's longevity.

13.1 Measuring Lead Pacing Thresholds

A temporary pacing mode, in which the pacing thresholds are measured, depends on the heart chamber in which the threshold is to be measured. In general, thresholds can be measured using a programmer in a comprehensive DDD mode, but for the sake of good visibility on a surface ECG record, it is recommended that a single-chamber mode be used; that is, AAI or A00 for the right atrium and VVI or V00 for the right and left ventricles. Optimally, a temporary LRL for measuring threshold must be set approximately 20/min higher than the patient's actual intrinsic heart rate; for that reason, the difference between asynchronous or inhibition modes should not be of any significance. For the sake of safety, the inhibition mode should be used because of the possible occurrence of ventricular extrasystoles. Usually, a voltage pacing threshold is measured while gradually decreasing the amplitude at a

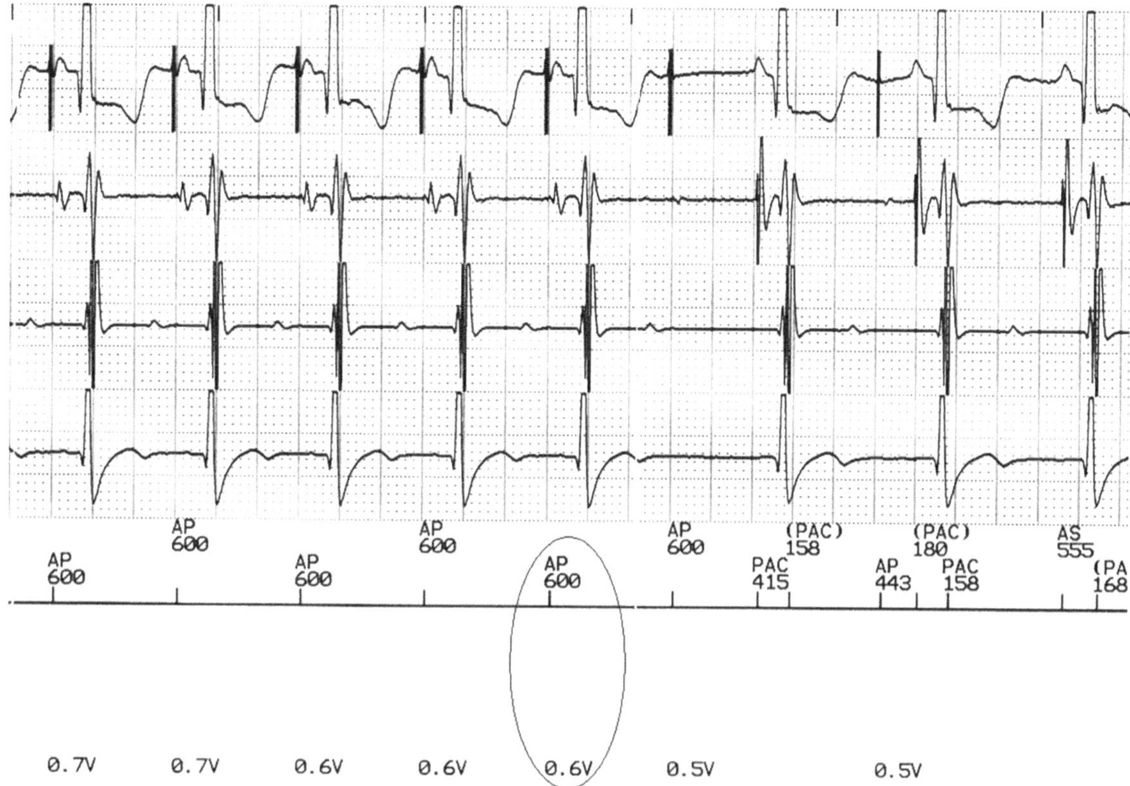

Fig. 13.2 Measurement of the right atrium voltage threshold in AAI mode

constant pacing pulse width. The monitored quantity values are decreased automatically by the implanted device, which is controlled by the programmer. The first unsuccessful or the last successful pacing pulse is identified, which represents the voltage pacing threshold. In addition, a pacing pulse width threshold can be measured at a constant voltage; this method is, however, not in use today.

13.1.1 Measuring the Threshold in the Right Atrium

In case of intact AV conduction, the AAI or A00 mode is to be applied temporarily to measure the threshold in the right atrium. Nevertheless, if a higher AV block (second or third degree) is diagnosed, measuring in the atrial mode could be dangerous because pulses could fail to be conducted to ventricles, which would result in a cardiac arrest. In such cases, the atrial threshold must be measured in the DDD mode. In the atrial mode (Fig. 13.2), the first unsuccessful atrial pacing pulse is identified while the amplitude is decreased gradually. Provided that a high pacing threshold is measured, certain pacing systems allow unipolar/bipolar configuration reprogramming; ICDs, however, lack this option.

In the DDD mode, the first unsuccessful atrial pacing pulse may not be apparent. Thus, prolongation of the cardiac cycle and an AV interval or changes in morphology are monitored, even though identification of these may also be insufficient if a less detailed ECG record lacking time markers is used.

13.1.2 Measuring the Threshold in the Right Ventricle

The VVI or V00 modes are applied advantageously for measuring threshold in the right ventricle. On a surface ECG, ventricular depolarization is usually more apparent than atrial depolarization. If the surface ECG is not used, shock lead electrograms may be used in ICD systems because there is a good signal due to a larger distance between the shock electrodes (Fig. 13.3). Provided that a high pacing threshold is measured in the right ventricle, certain pacing systems may allow unipolar/bipolar configuration reprogramming; ICDs, however, lack this option.

13.1.3 Measuring the Threshold in the Left Ventricle

The VVI or V00 modes also are applied for the purpose of measuring threshold in the left ventricle (Fig. 13.4). Provided that a high pacing threshold is measured in the left ventricle,

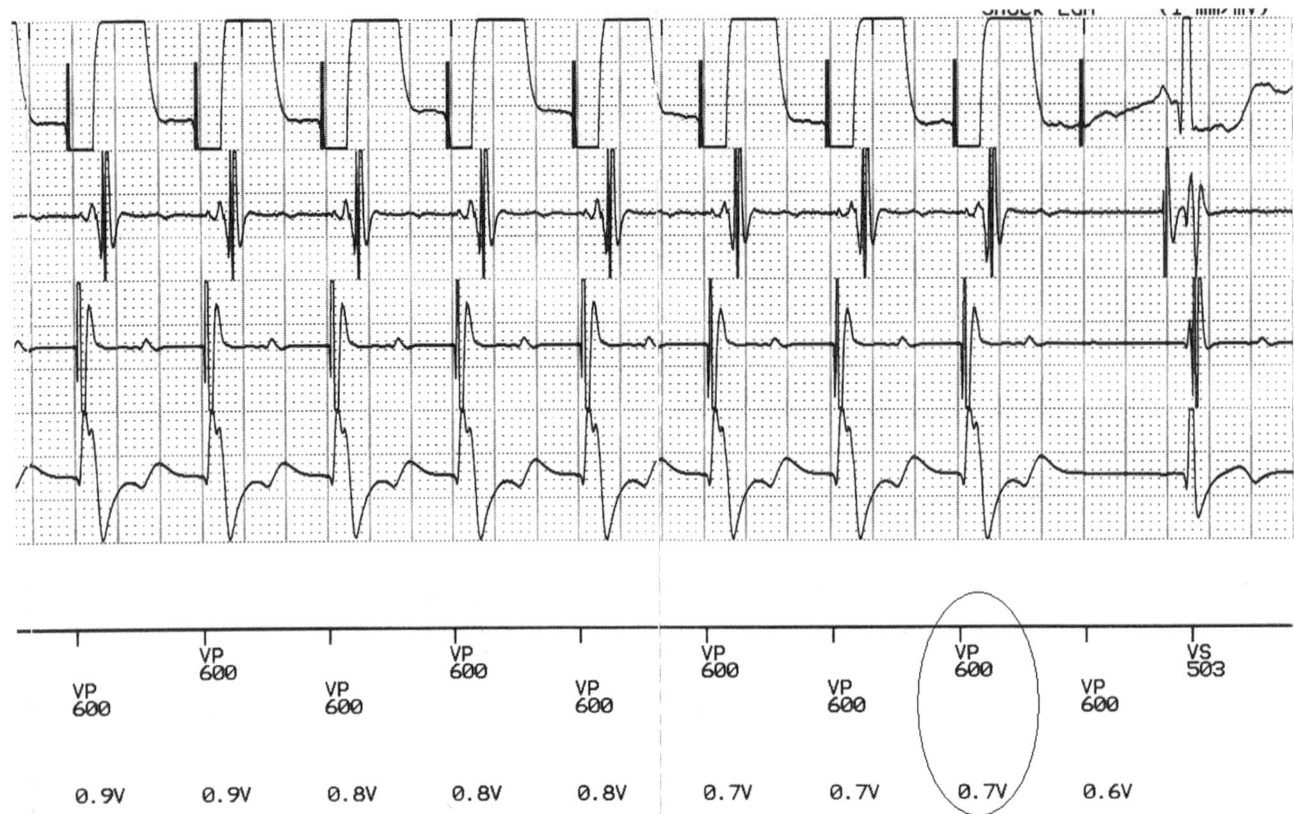

Fig. 13.3 Measurement of the right ventricle voltage threshold in VVI mode

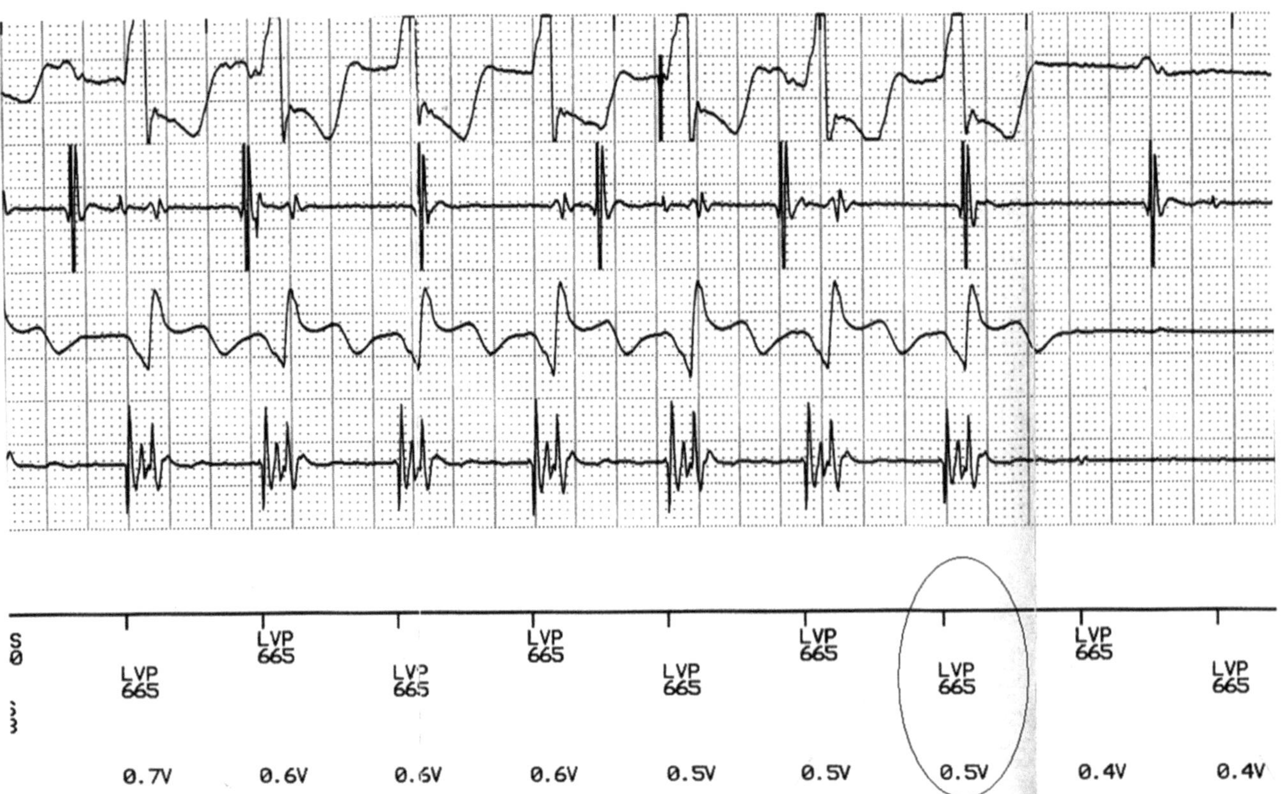

Fig. 13.4 Measurement of the left ventricle voltage threshold in VVI mode

certain pacing and defibrillation systems may allow reprogramming both to unipolar and bipolar configurations, as well as to an extended bipolar configuration with the use of the proximal electrode of the right ventricular lead.

13.2 Remote Patient Monitoring

First, it must be noted that the idea of remotely monitoring implanted devices is not novel. As early as at the beginning of 1970s, a concept of pacemaker longevity, transtelephonic monitoring (TTM), was introduced [85]. At the turn of the 1980s, the diagnostic features of the TTM system were extended by additional functions, including the sensing of intrinsic potentials, pacing successfulness, lead defects, and arrhythmias [86, 87].

Nevertheless, TTM pacemakers, commonly used in the USA, did not become massively popular in Europe. This method also required active cooperation on the part of the patient, which prevented mass spread of the use of this pacemaker [88].

At present, four systems operate or are about to be introduced globally. The launch has been delayed because of legislative issues concerning the protection of patient data or the use of a frequency spectrum. All systems work on a similar principle of a patient unit, communicating with an implanted device by means of various technical methods, and central data administration, that is, a telephonic or control computer center on a global or national level.

An implanted device, which may or may not be equipped with a transmitting antenna installed in the device header, transmits required diagnostic data to the patient unit. The transmission usually occurs in the 402- to 405-MHz frequency band, which is referred to as medical implant communication service, or in other country-specific frequency bands. Certain systems (Table 13.1) may also use a standard telemetry wand to communicate with the implanted device. The patient unit is connected to the management center via a telephonic or data network; the center receives and evaluates the data and distributes it to nursing staff or lower-tier medical centers. The data are transmitted on a regular basis, usually once a day; urgent information (such as a decrease in battery energy, detection of a severe arrhythmia, etc.) is transmitted immediately [89].

13.2.1 Home Monitoring™ (Biotronik, Germany)

Biotronik is a pioneering company in the field of remote pacemaker/ICD monitoring; the company started its operation in 2001. The CardioMessenger® patient unit is only slightly larger in size than a mobile phone and is capable of wireless communication with an implanted device within 2 m. The patient unit forwards obtained data to the management center using a GSM network; this is the main advantage of the system because mobile phone connections are more common than traditional dial-up connections today. This also means that a patient is monitored at all times; he or she can take the CardioMessenger device anywhere and is able to recharge it. The system is compatible with most available GSM networks. Reporting parameters may be set on a secured web site, which makes a visit to the physician unnecessary. Intracardial ECG records 30 s long are sent for the purpose of evaluation. Regular measurements are sent once a day at a preprogrammed hour set by the physician at a routine follow-up using a programming device supplied by the manufacturer. It is convenient to set an hour at night, while the patient is asleep and the patient unit is on the bedside table [89]. Currently, this system is the only system to offer monitoring of pacemakers, too, that is, it does not monitor only ICDs.

13.2.2 Latitude Patient Management System™ (Boston Scientific, USA)

This system was launched in Europe in 2009. The patient unit makes use of an analog line for the transmission of data to the management center. The line has to be configured specifically for the appropriate country. The possibility of connecting a wireless scale and a blood pressure gauge to the system for the purpose of monitoring progression of heart failure makes the system unique. The patient may also enter heart failure symptoms, such as fatigue, shortness of breath, and swelling, in the system on a weekly basis. Event reporting may also be set individually. Furthermore, the system allows optional transmission of data to various physicians, which improves communication concerning heart failure treatment [90].

Table 13.1 Overview of basic remote patient monitoring systems

	Home Monitoring™	Latitude™	CareLink™	Merlin.net™
Patient unit	Mobile	Fix	Fix	Fix
Data transfer to patient unit	MICS	ISM	MICS	MICS
Data transfer to center	GSM	Dial-up	Dial-up	Dial-up
Data transfer to center interval	Daily, serious events	Scheduled, serious events	Scheduled, serious events	Scheduled, serious events
Physician signalization	SMS, e-mail, fax	Fax, phone	SMS, e-mail	Fax, e-mail, SMS
Patient signalization	LED	Text or voice messages	LED	LED

MICS medical implant communication service *LED* light emitting diode *ISM* industrial, scientific, medical

13.2.3 CareLink Network™ (Medtronic, USA)

The CareLink system was launched in the USA in 2002; in Europe it has been available since 2007. The ICD communicates with a patient unit within 3 m. Data are sent via an analog telephonic line that is usable only in the patient's home country. Upon the occurrence of a more serious event, for example, an abnormal lead impedance, communication between the ICD and the patient unit is immediately established; failing that, attempts to establish communication are repeated at an interval of 3 h for a period of 3 days; subsequently, sound signals are emitted. The remote monitoring parameters may be set individually with a preset degree of severity for each patient. It is advantageous that the latest generations of devices are capable of conducting a wide range of measurements automatically, including automatic measurement of the threshold in the atrium and the right and left ventricles, which makes full remote control of a patient possible [90].

13.2.4 Merlin.net™ (St. Jude Medical, USA)

The Merlin.net wireless patient unit was introduced in 2008. The system communicates automatically with an implanted device and sends data to the physician via an analog telephone line; the use of a mobile network is planned. The possibility of signaling the results of a scheduled follow-up or remote follow-up on the patient unit is a useful function. The next generation of ICDs will measure pacing thresholds of all leads, which will make complete remote patient monitoring possible [90].

Currently, numerous clinical studies are being carried out with an aim to prove operational safety and the possibility of reducing routine follow-ups, be they scheduled or unscheduled (e.g., in the case of a defibrillation shock or suspicion of improper functioning of the implanted system). The main expected advantage of the monitoring systems being introduced is the reduction in the number of follow-ups along with the possibility to continuously monitor the patient's condition. Their importance even grows for patients with biventricular systems because increased demands are laid on pacing and, consequently, monitoring of implanted lead parameters.

Methods of payment for the introduction and administration of these systems and the protection of sensitive medical data from abuse has not yet been solved and can be seen as problematic. For the time being, the cost of implantation and operation of the system is included in the cost of the implantable devices. It may be assumed, however, that the issue will have to be dealt with in a systemic manner; this also applies to the method of analysis and dealing with sudden events in a patient. Today, physicians are overburdened by recordings from monitoring systems and cannot pay sufficient attention to the recordings; this will become an even more serious problem as the expected number of devices and monitored product categories increases.

14 Electromagnetic Compatibility and Technical Requirements

In various environments in which patients might find themselves (home, means of transport, a medical center, a workplace), electromagnetic fields may occur. The external fields may affect electronic devices that pacing technology primarily makes use of, and this phenomenon is called electromagnetic interference (EMI). Directives on medical devices require active implantable medical devices to be designed and produced so that EMI-related risks are eliminated or minimized.

14.1 The Electromagnetic Field Effects on Pacing Technology

Electromagnetic fields are characterized mainly by a dominant frequency and intensity. Their intensity decreases with increasing distance from a radiation source. Such a decrease in intensity depends on whether a person is in a near or far radiation field of the source. In most cases, people find themselves in the far-field; however, on certain occasions, it may be the opposite. The distance from the source where the near-field passes into the far-field depends on the relationship between the size of the field source and the radiation wavelength. In the near-field, the intensity is, as a rule, inversely proportional to the third power of the distance; in the far-field, the intensity is inversely proportional to the first power of the distance. In fact, the situation is more complex –for example, the source geometry, whether the electric or magnetic component of the electromagnetic field prevails also matter. In general, then, the electromagnetic susceptibility of pacing technology depends on maximum field amplitudes rather than effective values.

Even in people without implanted active implantable medical devices (AIMDs), the external electromagnetic field induces currents in body tissue. Depending on the frequency, the external field may stimulate nerves, damage cell membranes, or heat tissues. At frequencies below 100 kHz, nerves may be paced at lower levels of the field. The effects on nerve stimulation decrease with the frequency and have an immediate effect. At these frequencies, the human body is almost transparent for the magnetic component of the field; magnetic field of internal tissue is thus level with external magnetic field. At low frequencies (below 100 kHz), the electric component is much weaker than the external field because tissue is conductible, and the human body tissue attenuates electromagnetic fields. Starting from frequencies of around 5 MHz, body tissue attenuates electromagnetic fields. At frequencies over 10 MHz, the tissue is heated at lower levels. Contrary to nerve stimulation, heating is an accumulative process. For frequencies ranging from 100 kHz to 10 MHz, both effects (stimulation and heating) may occur.

In people with implants, the field effects on body tissue may intensify. A metal implant may heat up and, as a consequence, heat the surrounding tissue. The field may also increase the current density in the body tissue around the device. To eliminate clinically significant effects, the density of the induced current must be reduced.

Heart tissues are connected to pacing devices by leads. As far as electric and magnetic components are concerned, the inductive methods and voltage values in the leads depend on the configuration (unipolar or bipolar) of the applied pacing and dimensions. The bipolar configuration is far less sensitive to interference than the unipolar configuration. The field evaluation criterion is thus based on the EMI-sensitivity of pacemakers with unipolar leads. Bipolar lead systems (defibrillators and the majority of pacemakers) are less sensitive to interference. An external electromagnetic field may induce voltage and currents in the lead, which will be sensed by the device. In stronger fields, these signals will be evaluated as an intrinsic cardiac activity and will result in inhibition of pacing or detection of tachycardia. Moreover, pacing devices make use of inductive coupling or a radio connection to communicate with a programmer. If the frequencies of external electromagnetic fields are close to those used for communication with the programmer, the communication may be interfered with or disrupted. In such cases, the programmer must be placed further from electrical devices and cables must not be crossed.

D. Korpas, *Implantable Cardiac Devices Technology*, DOI 10.1007/978-1-4614-6907-0_14,

In pacemakers or in terms of the ICD pacing function, a strong EMI may result in the following situations [17]:

- Pacing inhibition
- Switching to asynchronous pacing
- Tracking the interfering rhythm to ventricle pacing
- Lead system–induced current
- Switching of a magnetic switch
- Heating of pacing leads
- Damage to device electronics

In ICDs, the following situations may also occur [72]:

- Switching off tachycardia therapy
- Undesirable shock as a consequence of the detection of an interfering signal as tachycardia

The danger of EMI effects on a patient depends on the period of exposure, distance from the source, and the position of a patient or implanted system in reference to the field source. It is almost impossible to estimate in advance the possible effects of EMI on the operation of pacing technology.

Possible significant EMI sources include:

- High voltage lines
- Radio communication devices (transmitters, amplifiers, radar)
- Arc welders
- Electric induction devices
- Electric manual tools
- Electronic article surveillance systems
- Electric furnaces
- Vehicle alternators

14.2 Technical Standardization and Pacing Technology Tests

A technical standard is a documented requirement concerning operational, functional, or performance properties, promulgated by an official body, which must be met by products or services for an intended purpose. Compliance with the technical standards can be mandatory or voluntary, depending on the country. This has changed during the history of implanted devices. Nevertheless, standards establish the reference values and safety criteria or are required in public tenders, and compliance with the technical standards can affect the rollout of a product (declaration of conformity).

The basic technical standards covering implantable pacing technology are international standards in the International Organization for Standardization (ISO) 14708 series:

- ISO 14708-1, Implants for surgery – Active implantable medical devices – Part 1: General requirements for safety, marking and information to be provided by the manufacturer
- ISO 14708-2, Implants for surgery – Active implantable medical devices – Part 2: Cardiac pacemakers
- ISO 14708-6, Implants for surgery – Active implantable medical devices – Part 6: Particular requirements for active implantable medical devices intended to treat tachycardia (including implantable defibrillators) and the European standards EN 45502 series:
- EN 45502-1, Active implantable medical devices – Part 1: General requirements for safety, marking, and information to be provided by the manufacturer
- EN 45502-2-1, Active implantable medical devices – Part 2-1: Particular requirements for active implantable medical devices intended to treat bradycardia (cardiac pacemakers) [17]
- EN 45502-2-2, Active implantable medical devices – Part 2-2: Particular requirements for active implantable medical devices intended to treat tachycardia (includes implantable defibrillators) [72]

Corresponding standards in the ISO and EN systems are practically the same and establish the same technical requirements and reference values for testing the safety of implantable devices and their components. All designated pacing technology tests are specified for devices at 37±2 or 37±5 °C. Because established designs are not temperature sensitive within such a temperature range, this is believed to be sufficient to validate an implantable device at thermal equilibrium after implantation.

14.2.1 Packaging, Leakage Currents, Dissipating Heat, and Power Source Requirements

The outer packaging of implantable system components shall contain information and parameter values for the purpose of complete identification, and a notification that ICDs are always supplied with inactivated tachycardia treatment. Furthermore, the designation of connector types and dimensions must also be included so that compatible system components may be identified without having to open the sterile packaging. As far as pacemakers are concerned, information on the most comprehensive pacing mode available and the mode set on dispatch must be provided. On ICD packaging, available tachycardia therapies must be listed, and a warning of the necessity of inactivating the tachycardia treatment during surgical interventions must be given.

The accompanying documentation must contain information on the average estimated longevity (in years) of the device. The longevity should depend on the total number of maximum-energy shocks. All packages containing sterile material must carry information on the expiration date in the year-month-day sequence. Provided that expected application of an AIMD implantable component closed in sales packaging requires the component to be connected to other devices or accessories not contained in the packaging, the sales packaging must specify the connector type (pacing/sensing, cardioversion/defibrillation), configuration (unipolar/bipolar), and connector geometry (length and diameters in millimeters or a reference to issued standards).

Each lead, and possibly each adaptor, must be permanently and visibly marked with the manufacturer and type designation and a serial number or a lot number. Naturally, it is required that any implantable system component is sterile upon unpacking and contains no excessively loosened solid particles – sterile dust. The amount of particles depends on the surface area, not the lead content. This particularly applies to leads typically having a large surface area but small contents. Reference values are based on a standardized particulate contamination test. An implantable device must also be designed to resist minor mechanical impacts caused by manual handling during implantation.

Sustained direct currents leaking from implanted leads may result in damage to tissues or material corrosion. Unless on purpose, an implantable device must always be electrically neutral upon application. Direct leakage current exceeding 1 μA must not be induced in any current pathway between defibrillation lead connectors and the can, and direct leakage current exceeding 0.1 μA must not be induced in the current pathway of any pacing connector. A test for direct leakage current is carried out with inactivated tachycardia therapies. Load resistances are applied, which substitute impedance in an implanted device. Limits for pacing/sensing connectors comply with ISO 14708-2. In shock connectors, the limits are ten times higher as a result of the much larger area of the shock lead. In pacing/sensing leads with an area of pacing electrodes less than 10 mm^2, leakage current of approximately 10 μA is acceptable. In shock leads with the area exceeding 300 mm^2, maximum direct leakage current of approximately 100 μA is acceptable. Alternating leakage currents may occur while high-voltage ICD capacitors are being charged. Low-frequency currents may induce ventricular fibrillation, whereas high-frequency currents may cause heating and damage to tissues.

During charge and discharge cycles, a considerable amount of energy may dissipate inside an ICD. Today's ICDs, however, have sufficient weight, making them capable of dissipating the heat; thus the temperature will not increase by more than 4 °C at any point on the ICD's external surface. Possible tissue damage is a function of exposure time. Measured surface temperatures during testing shall be identical to temperatures measured in an implanted device. At present, 43 °C is considered the temperature threshold for tissue damage. Most damage to muscular and adipose tissue by temperatures ranging between 43 °C and 45 °C are reversible; under higher temperatures, necrosis occurs. Local chronic heating of tissue caused by an AIMD should be restricted to a maximum temperature of 41 °C.

The manufacturer must specify the time of recommended replacement of an implantable device. Because the ICD-provided treatment is of vital importance, it was necessary to determine a minimum number of maximum-energy shocks an ICD must be capable of applying after the recommended replacement time is identified. Considering the period of time between routine follow-ups, a minimum of six maximum-energy shocks can occur after the electric replacement indicator is activated.

14.2.2 Requirements for Implantable Leads

Implantable leads are subject to tests of both electrical and mechanical parameters. For example, defibrillation lead conductors must pass an axial load test, and, at the same time, the capability of delivering maximum defibrillation shocks must not be impaired. A damaged lead conductor may have a lesser ability to carry high currents, even though its impedance is within the appropriate limits. In the axial load test, a defibrillation shock of 1,000 V is simulated from a capacitor with a capacity of 200 μF at a system impedance between 20 and 25 Ω. Hence, all test values are overrated if compared with values in practice. Ten shocks are applied for the purpose of testing, and signs of damage by the test current then are searched for visually [72].

Minimum requirements for implantable lead resistance to bending should be determined by means of mechanical testing. A lead conductor or connector must withstand a minimum of 47,000 – or possibly 82,000 – load cycles flawlessly. The manufacturer determines the specimen size, data analysis, and safety reserve to meet the requirements for the minimum number of cycles. Furthermore, the manufacturer is responsible for defining a comprehensive set of requirements concerning the reliability of a specific lead conductor solution. The value of 5 N has been agreed upon as the endurance limit for axial tensile force exerted on an implanted lead. In particular, resistance to bending (fracture) and lead insulation resistance are tested. A larger bend radius naturally causes less mechanical stress and a smaller bend radius causes more stress. Thus the minimum number of bending cycles decreases with higher stress. According to standards, the tests usually are conducted on an accelerated basis; a higher load is exerted on leads on purpose, which results in shorter longevity of test specimens.

The lead characteristics, as specified in documentation, must provide the following information:

- General description of materials used in individual components (a connector, insulation, a conductor, pacing and defibrillation electrodes)
- Notification of the drug content and the drug identification
- Nominal values of mechanical dimensions (length, geometric surface area of all electrodes, insertion diameter, distance between lead pacing electrodes)
- Electrical characteristics (conductor resistance, pacing impedance, sensing impedance)

Connectors of implantable devices, which connect the devices and leads, must be identified by their type. Connector retention force shall be ≥ 7.5 N [17, 72].

14.2.3 EMI Susceptibility Requirements

As mentioned earlier, external electromagnetic fields may induce currents flowing from leads to the heart, which may result in fibrillation or local overheating. Moreover, the fields may induce voltage in the lead conductor, potentially damaging the implantable device or preventing it from properly sensing the intracardial signal. Thus, implantable systems ought to have a certain susceptibility to electromagnetic interference.

Technical standards [17, 72] define test procedures and reference levels to protect implanted systems from:

- Damage or fibrillation caused by currents induced directly in implanted leads or conducted by interference currents from the device
- Permanent failure caused by voltages induced in implanted leads
- Unacceptable changes or operation modes caused by voltages induced in implanted leads
- Temporary changes in the device's therapeutic behavior caused by voltages induced in implanted leads
- Temporary changes in the device's therapeutic behavior caused by weak static magnetic fields (1 mT) affecting any magnetic-sensitive components of an implantable device
- Permanent failure caused by stronger static magnetic fields (50 mT) affecting any magnetic-sensitive components of an implantable device
- Persisting malfunction of the device caused by time-varying magnetic fields applied to an implantable device

EMI susceptibility tests are conducted in several frequency bands, ranging from 16.6 Hz to 3 GHz. Special attention is paid to certain frequencies used in power engineering. Authors of technical standards have agreed on a bipolar sensitivity of 0.3 mV as suitable for EMI testing with frequencies of more than 1 kHz. In patients with ICDs, sensing sensitivity set below 0.3 mV increases the risk of undesirable sensing of remote signals.

Permissible human exposure to an electromagnetic field is limited by a number of both national and international directives and recommendations. Previously, standards only applied to a limited range of frequencies (up to 30 MHz). Today, the standard also includes higher frequencies to cover contemporary telecommunication devices. Implanted leads function as antennas in the electromagnetic field. Voltages induced in a lead depend on the position and characteristics of the lead as well as frequency, polarization, and the orientation of the electromagnetic field. At low frequencies up to several megahertz, each lead forms a conducting loop, where voltages are induced in proportion to the field frequency. The induction loop area is considerably larger in unipolar lead conductors than in bipolar conductors.

At higher frequencies, leads function as dipole antennas. In bipolar leads, voltages are induced between lead proximal and distal electrodes. EMI susceptibility tests are carried out by means of a voltage clamp. The frequency range of the test voltages is wider than the frequency range of physiological signals, and as the voltage level grows, an implantable device may start detecting interference at a certain level. This may affect the therapeutic behavior of the implantable system (e.g., pacing inhibition, tachycardia detection). More test signals are applied, taking into consideration various interferences a person may be exposed to at public places.

Per the ANSI/AAMI PC69 standard, a procedure is applied to testing radiation susceptibility with use of mobile phones. During this test, a device together with all leads is immersed into saline solution, simulating body tissue, and is exposed to a near-field of an electric dipole. According to ANSI/AAMI PC69, two field levels are applied for testing. The lower level tests the system's function upon exposure to a mobile phone with 2 W output at a distance of 15 cm. Passing the test is mandatory. The higher level tests the exposure in the distance of 2 cm, which simulates a mobile phone placed above an implanted device. Passing the test is not mandatory yet. Per ANSI/AAMI PC69, the test signal is modulated. This test also guarantees susceptibility in the far-field of powerful transmitters, such as mobile phone base stations.

As far as protection from magnetic fields is concerned, the device must be susceptible to exposure to weak magnetic fields up to the value of magnetic induction 1 mT to prevent activating the magnetic switch. Nevertheless, strong magnetic fields (more than 50 mT) may interfere with the operation of an implantable medical device. The magnetic switch may be activated and telemetry may be interfered with or a therapy may be suppressed. The manufacturer must also assess the risk arising from permanent activation of the magnetic switch. The tests usually are conducted up to the effective value of the magnetic field intensity of 150 A/m [17, 72].

14.3 Ionizing Radiation Effects

The operation of pacing devices may also be affected by ionizing radiation. In addition to possible EMI by ionizing radiation sources (e.g., radiology devices), ionizing radiation may effect the operation of device electronics and the battery. Recent devices making use of complementary metal oxide semiconductor technology are, however, much less sensitive to the ionizing radiation. Based on the results of studies prepared by the American Association of Physicists in Medicine, recommendations were issued in 1994; these are usually included in the manufacturer's documentation but are not generally known.

In a silicon semiconductor, ionizing radiation excites electron into what is referred to as a conduction band. If the semiconductor is in an electrical field, the conductibility increases. Interactions of neutrons with silicon nuclei are significant. Upon collision of a fast neutron with a silicon atom, the atom is forced from its lattice position, and the silicon atoms thus make imperfections in the crystal lattice [91, 92].

For patients with pacing or defibrillation systems, radiology diagnostic or therapeutic devices in medical centers are usually the most common sources of ionizing radiation. The ionizing radiation effects are influenced by several factors, including, among others, the distance between a radiation beam and the device, the beam type and energy, the dose rate, the total dose received over the period of the device longevity, and the device shielding. The effects of ionizing radiation may also differ depending on individual types of devices and manufacturers. During ionizing radiation treatment (linear accelerators, brachytherapy, and betatrons), it is recommended that the device be shielded by a protective shield no matter its distance from the radiation beam. The center of the beam should not point directly at the device. After therapeutic irradiation, the function of the device, including sensing, pacing thresholds, and capacitor re-formation, must be checked. The majority of diagnostic display methods, for example, X-ray skiagraphy and fluoroscopy or computed tomography (CT), are not considered as dangerous. If a patient is examined using CT and the device is not directly in the beam of the CT image, it is not affected. If it is placed directly in the main CT pencil beam, excessive sensing may occur throughout the period of exposure to the beam.

If an implanted device is exposed to intensive ionizing radiation or other extreme conditions, the device may be reset or switched into a safety mode, which devices are equipped with today. As a consequence, the device is switched into an operation mode with basic parameters that are considered safe for the majority of patients [93].

Because of stochastic effects, it is impossible to determine a safe radiation dose or guarantee the proper function of a device upon exposure to ionizing radiation. Recommendations specify a maximum recommended safe dose of 2 Gy. Experiments dealing with monitoring the in vitro operation of implantable systems irradiated by ionizing radiation from a linear accelerator proved the operation, even under considerably higher repeated doses; however, the patient's safety must always be the first concern.

References

1. Schecter DC (1971) Early experience with resuscitation by means of electricity. Surgery 69(3):360–372
2. Althaus J (1864) Report of the committee appointed by the Royal Medical and Chirurgical Society to inquire into the uses and the physiological, therapeutical and toxical effects of chloroform. Med Chir Trans 47:416
3. Baxi MV (2003) First artificial pacemaker: a milestone in history of cardiac electrostimulation. Asian Stud Med J. http://asmj.netfirms.com/article0903.html. Accessed 18 Mar 2011
4. Vrana M et al (1984) Electronic devices for organs and tissue stimulation, 1st edn. SNTL, Prague
5. Zeman K (2009) Heart conduction system electrophysiology in time helix. Kardiol Rev 11(4):164–165
6. Blair HA (1936) On the quantity of electricity and the energy in electrical stimulation. J Gen Physiol 19(6):951–964, Jul 20
7. Bauwens P (1941) Electro-diagnosis and electrotherapy in peripheral nerve lesions: (section of physical medicine). Proc R Soc Med 34(8):459–468
8. Hyman AS (1969) Resuscitation of the stopped heart by intracardiac therapy. II. Experimental use of artificial pacemaker. Arch Intern Med 50:283
9. Larsson B, Elmqvist H, Ryden L, Schuller H (2003) Lessons from the first patient with an implanted pacemaker: 1958–2001. Pacing Clin Electrophysiol 6:114–124
10. Lipoldova J, Novak M (2006) History of permanent pacing. Kardiol Rev 8(4):166–173
11. Lüderitz B (2002) History of the disorders of cardiac rhythm, 3rd edn. Futura, New York
12. Special Committee on Aging US Senate (1982) Fraud, waste, and abuse in the medicare pacemaker industry. U.S. Government printing office, Washington
13. CCC del Uruguay: historical CCC's pacemakers. http://www.ccc.com.uy/pacemaker/index.htm. Accessed 22 July 2012
14. The Neurostimulation Technology Portal: history of pacemakers. http://www.biotele.com/pacemakers.htm. Accessed 14 July 2012
15. Jeffrey K, Parsonnet V (1998) Cardiac pacing, 1960–1985: a quarter century of medical and industrial innovation. Circulation 97(19):1978–1991, May 19
16. Medtronic Corporation: Adapta, Versa, Sensia – programming guide (EE). www.medtronic.com/manuals. Accessed 30 Sept 2010
17. European Committee for Electrotechnical Standardization (2003) CENELEC EN 45502-2-1: Active implantable medical devices part 2-1: particular requirements for active implantable medical devices intended to treat bradyarrhythmia (cardiac pacemakers), 1 Dec 2003
18. Altrua 50 and Altrua 60 system guide. Boston Scientific, St. Paul, © 2008, 278 p
19. Medtronic Corporation: Advisa DR MRI™ SURESCAN™ A3DR01 – clinician manual. www.medtronic.com/manuals. Accessed 30 Sept 2010
20. Vymazal J, Taborsky M, Zacek R (2009) Magnetic resonance imaging with implanted cardiac pacemaker EnRhythm MRI SureScan with MR compatible electrodes CapSureFix MRI – first experience. Cesk Radiol 63(1):9–12
21. Cihak R (1997) Anatomy 3, 1st edn. Grada, Prague, 655 p
22. Trojan S (2003) Medical physiology, 4th edn. Grada, Prague, 771 p
23. Bytesnik J, Cihak R (1999) Arrhythmias in medical praxis, 1st edn. Triton, Prague, 179 p
24. Cihalik C (1998) An atlas of clinical electrocardiography, 1st edn. Palacky University, Olomouc, 215 p
25. ECG Education webpage. http://ekg.kvalitne.cz. Accessed 11 Dec 2010
26. Lukl J, Heinc P et al (2001) Modern treatment of arrhythmias, 1st edn. Grada, Prague, 212 p
27. Lukl J et al (1997) Pokroky v arytmologii, 1st edn. Grada, Prague, 210 p
28. Gandalovicova J (2002) Chronic heart failure ventricular arrhythmias. Intern Med Prax 4:181–187
29. Heinc P (2005) Practical aspects and news in pharmacological treatment of arrhythmias. Kardiol Rev 7(1):10–16
30. Lukl J (2005) Pharmacotherapy arrhythmias. Intern Med Prax 3:115–119
31. Bernstein AD, Daubert J-C, Fletcher RD, Hayes DL, Lüderitz B, Reynolds DW, Schoenfeld MH, Sutton R (2000) The revised NASPE/BPEG generic code for antibradycardia, adaptive-rate, and multisite pacing. PACE 25:260–264
32. Insignia AVT/Ultra system guide. Guidant Corporation, St. Paul, © 2003, 283 p
33. Tricoci P, Allen JM, Kramer JM, Califf RM, Smith SC (2009) Scientific evidence underlying the ACC/AHA clinical practice guidelines. JAMA 301(8):831–841
34. Epstein AE et al (2008) ACC/AHA/HRS 2008 guidelines for device-based therapy of cardiac rhythm abnormalities. J Am Coll Cardiol 51(21):e1–e62, May 27
35. Epstein AE et al (2008) ACC/AHA/HRS 2008 guidelines for device-based therapy of cardiac rhythm abnormalities: executive summary. Heart Rhythm 5(6):934–955
36. Priori SG, Aliot E, Blomstrom-Lundqvist C et al (2003) Update of the guidelines on sudden cardiac death of the European society of cardiology. Eur Heart J 24:13–15
37. Kautzner J et al (2007) Syncope: diagnosis and treatment. Supplementum Cor Vasa 49(11):63–73

D. Korpas, *Implantable Cardiac Devices Technology*, DOI 10.1007/978-1-4614-6907-0,

38. Duska F (2008) Medical publication reading and evaluation. http://www.lf3.cuni.cz/cs/pracoviste/imunologie/vyuka/pvk/. Accessed 14 Nov 2010
39. Bytesnik J (2002) The efficacy of implantable cardioverter-defibrillators in the prevention of sudden cardiac death: a survey of the current knowledge. Interv Akut Kardiol 1:32–36
40. Ellenbogen KA, Wood MA et al (2008) Cardiac pacing and ICDs, 5th edn. Blackwell, Malven, 558 p
41. Ozabalova E (2007) Cardiac resynchronization therapy for patients with chronic heart failure. Ph.D. thesis, Faculty of Medicine, MU Brno, 71 p
42. Bytesnik J (2004) "Hot information" from the area of primary prevention of sudden death: studies with implantable cardioverters-defibrillators DINAMIT and SCD-HEFT. Interv Akut Kardiol 3:139–141
43. Krivan L (2005) Primary prevention of sudden cardiac death. Intern Med Prax 1:7–9
44. Heart failure (HF): congestive heart failure. Merck manual professional. http://www.merckmanuals.com/professional/cardiovascular_disorders/heart_failure/heart_failure_hf.html. Accessed 11 Jan 2011
45. Svab P, Simon J (2003) Resynchronization therapy for chronic heart failure – summary of current knowledge. Intern Med Prax 1:14–18
46. Neuzil P, Taborsky M, Sediva L (2004) Prognostic impact of cardiac resynchronization therapy. Cardiology revue – extra ed, 6:39–42
47. Kautzner J (2002) Biventricular pacing: a novel therapeutic option in management of chronic heart failure. Intern Med Prax 6:273–278
48. Acuity steerable physician manual. Guidant Corporation, St. Paul, © 2005, 40 p
49. Flextend 2 physician manual. Boston Scientific, St. Paul, © 2009, 28 p
50. Technical manual Selox SR. Biotronik GmbH & Co., Berlin, © 2004, 35 p
51. Endotak Reliance G, Endotak Reliance SG physician manual. Guidant Corporation, St. Paul, © 2006, 40 p
52. (2010) Technical manual ICD leads. Biotronik SE & Co. KG, Berlin, 113 p
53. Bulava A, Lukl J (2005) Resynchronization therapy in the treatment of congestive heart disease – review of literature. Intern Med Prax 5:229–236
54. Kautzner J, Riedlbauchova L, Cihak R, Bytesnik J, Vancura V (2004) Technical aspects of implantation of LV lead for cardiac resynchronization therapy in chronic heart failure. Pacing Clin Electrophysiol 27(6 Pt 1):783–790
55. Medtronic Corporation: Medtronic Attain bipolar OTW 4194 Technical manual. www.medtronic.com/manuals. Accessed 23 Nov 2010
56. Easytrak 2 IS-1 Physician manual. Guidant Corporation, St. Paul, © 2006, 28 p
57. (2010) Technical manual Corox OTW-L BP, Corox OTW BP, Corox OTW-S BP. Biotronik SE & Co. KG, Berlin, 54 p
58. Medtronic Corporation: Medtronic Capsure EPI 4965 Technical manual. www.medtronic.com/manuals Accessed 25 Nov 2010
59. Physician manual Lead tunneller. Boston Scientific, St. Paul, © 2009, 20 p
60. The Boston Scientific CRM Compass. Boston Scientific, St. Paul, © 2007, 216 p
61. Reference guide to pacemakers, ICDs and leads. Boston Scientific, St. Paul, © 2007, 270 p
62. Medtronic Corporation: Adapta DR – Implant manual. www.medtronic.com/manuals. Accessed 30 Sept 2010
63. Medtronic Corporation: Adapta, Versa, Sensia – Reference guide (EE). www.medtronic.com/manuals. Accessed 30 Sept 2010
64. Novak M, Psenicka M, Smola M, Briza J, Cermak S (1997) Dual-chamber rate-regulated cardiac pacing is aiming toward automatic pacemakers. Bratisl Lek Listy 98(11):604–608
65. Medtronic Corporation: Concerto® II CRT-D D294TRK – Clinician manual (EE). www.medtronic.com/manuals. Accessed 30 Sept 2010
66. (2007) Cylos technical manual. Biotronik GmbH & Co., Berlin, 138 p
67. (2009) Talos technical manual. Biotronik GmbH & Co., Berlin, 48 p
68. Cardiac rhythm management division U.S. product catalog. St. Jude Medical, St. Paul, © 2010, 71 p
69. Barold S, Stroobandt XR, Sinnaeve FA (2004) Cardiac pacing step by step an illustrated guide, 1st edn. Blackwell, Malven, 341 p
70. Teligen 100 system guide. Boston Scientific, St. Paul. © 2009, 283 p
71. Bernstein A, Camm AJ, Fletcher RD et al (1993) North American society of pacing and electrophysiology policy statement. The NASPE/BPEG defibrillator code. Pacing Clin Electrophysiol 16(9):1776–1780
72. International Organization for Standardization (2010) ISO 14708-6:2010: Implants for surgery – active implantable medical devices – part 6: particular requirements for active implantable medical devices intended to treat tachyarrhythmia (including implantable defibrillators), 25 Feb 2010
73. Strydis Ch (2005). Implantable microelectronic devices. M.Sc. thesis. http://ce.et.tudelft.nl/publicationfiles/1038_555_thesis_strydis.pdf. Accessed 6 Jan 2011
74. Medtronic Corporation: Maximo DR 7278 – Implant manual. www.medtronic.com/manuals. Accessed 30 Sept 2010
75. See Ref. [66]
76. Theuns DAMJ, Rivero-Ayerza M, Goedhart MD, Miltenburg M, Jordaens JL (2008) Morphology discrimination in implantable cardioverter-defibrillators: consistency of template match percentage during atrial tachyarrhythmias at different heart rates. Europace 10:1060–1066
77. Medtronic Corporation: Protecta CRT-D D364TRG – Clinician manual (EE). www.medtronic.com/manuals. Accessed 24 Nov 2010
78. Mocanu D et al (2004) Patient-specific computational analysis of transvenous defibrillation: a comparison to clinical metrics in humans. Ann Biomed Eng 32(6):775–783
79. Jacob S et al (2010) High defibrillation threshold: the science, signs and solutions. Indian Pacing Electrophysiol J 10(1):21–39
80. Bax JJ et al (2005) Cardiac resynchronization therapy part 1-issues before device implantation. JACC 46(12):2153–2167
81. Bax JJ et al (2005) Cardiac resynchronization therapy part 2-issues during and after device implantation and unresolved questions. JACC 46(12):2168–2182
82. System guide Cognis 100-D CRT-D. Boston Scientific, St. Paul, © 2009, 306 p
83. Crawford MH et al (1999) ACC/AHA guidelines for ambulatory electrocardiography-part III. JACC 34(3):912–948
84. Guide to replacement of cardiac implantable pulse generators. Boston Scientific, St. Paul, © 2007, 33 p
85. Furman S, Escher DJW (1975) Transtelephone pacemaker monitoring. In: Schaldach M, Furman S (eds) Advances in pacemaker technology. Springer, Berlin, pp 177–194
86. Griffin JC, Schuenemeyer TD, Hess KR (1986) Pacemaker follow-up: its role in the detection and correction of pacemaker system malfunction. Pacing Clin Electrophysiol 9:387–391
87. Vallario LE, Leman RB, Gillette PC (1988) Pacemaker follow-up and adequacy of Medicare guidelines. Am Heart J 116:11–15
88. Theuns DA, Res JC, Jordaens LJ (2003) Home monitoring in ICD therapy: future perspectives. Europace 5(2):139–142
89. Lipoldova J (2009) Remote monitoring of patients with implanted cardiac pacemaker. http://universitas.muni.cz/200704/DMPsPM.htm. Accessed 15 Oct 2009
90. Burri H, Senouf D (2009) Remote monitoring and follow-up of pacemakers and implantable cardioverter defibrillators. Europace 11(6):701–709
91. Musilek L (1992) Dosimetry of ionizing radiation (inclusive methods). FJFI CVUT, Prague
92. Penhaker M, Kobza F, Tiefenbach P (2007) Czech – English dictionary for thematic biomedical fields, 1st edn. VŠB-TU, Ostrava, 64 p
93. Medtronic Kappa™ 400 series pacemaker information and programming guide. Medtronic, Inc. © 1997, 340 p
94. Love JC (2006) Cardiac pacemakers and defibrillators, 2nd edn. Landes Bioscience, Georgetown, 174 p

Index

D. Korpas, *Implantable Cardiac Devices Technology*, DOI 10.1007/978-1-4614-6907-0,

MIX
Papier aus verantwortungsvollen Quellen
Paper from responsible sources
FSC® C105338

If you have any concerns about our products,
you can contact us on
ProductSafety@springernature.com

In case Publisher is established outside the EU,
the EU authorized representative is:
Springer Nature Customer Service Center GmbH
Europaplatz 3, 69115 Heidelberg, Germany

Printed by Libri Plureos GmbH
in Hamburg, Germany